Advances in Intelligent Systems and Computing

Volume 1336

The series "Advances in Intelligent Systems and Computing" contains publications on theory, applications, and design methods of Intelligent Systems and Intelligent Computing. Virtually all disciplines such as engineering, natural sciences, computer and information science, ICT, economics, business, e-commerce, environment, healthcare, life science are covered. The list of topics spans all the areas of modern intelligent systems and computing such as: computational intelligence, soft computing including neural networks, fuzzy systems, evolutionary computing and the fusion of these paradigms, social intelligence, ambient intelligence, computational neuroscience, artificial life, virtual worlds and society, cognitive science and systems, Perception and Vision, DNA and immune based systems, self-organizing and adaptive systems, e-Learning and teaching, human-centered and human-centric computing, recommender systems, intelligent control, robotics and mechatronics including human-machine teaming, knowledge-based paradigms, learning paradigms, machine ethics, intelligent data analysis, knowledge management, intelligent agents, intelligent decision making and support, intelligent network security, trust management, interactive entertainment, Web intelligence and multimedia.

The publications within "Advances in Intelligent Systems and Computing" are primarily proceedings of important conferences, symposia and congresses. They cover significant recent developments in the field, both of a foundational and applicable character. An important characteristic feature of the series is the short publication time and world-wide distribution. This permits a rapid and broad dissemination of research results.

Indexed by SCOPUS, DBLP, EI Compendex, INSPEC, WTI Frankfurt eG, zbMATH, Japanese Science and Technology Agency (JST), SCImago.

All books published in the series are submitted for consideration in Web of Science.

More information about this series at http://www.springer.com/series/11156

Arkadiusz Mężyk · Sławomir Kciuk ·
Roman Szewczyk · Sławomir Duda
Editors

Modelling in Engineering 2020: Applied Mechanics

 Springer

Editors
Arkadiusz Mężyk
Department of Theoretical
and Applied Mechanics
Faculty of Mechanical Engineering
The Silesian University of Technology
Gliwice, Poland

Sławomir Kciuk
Department of Theoretical
and Applied Mechanics
Faculty of Mechanical Engineering
The Silesian University of Technology
Gliwice, Poland

Roman Szewczyk
Institute of Metrology
and Biomedical Engineering
Faculty of Mechatronics
Warsaw University of Technology
Warsaw, Poland

Sławomir Duda
Department of Theoretical
and Applied Mechanics
Faculty of Mechanical Engineering
The Silesian University of Technology
Gliwice, Poland

ISSN 2194-5357 ISSN 2194-5365 (electronic)
Advances in Intelligent Systems and Computing
ISBN 978-3-030-68454-9 ISBN 978-3-030-68455-6 (eBook)
https://doi.org/10.1007/978-3-030-68455-6

This Springer imprint is published by the registered company Springer Nature Switzerland AG
The registered company address is: Gewerbestrasse 11, 6330 Cham, Switzerland

Foreword

Advances in Intelligent Systems and Computing—Modelling in Engineering 2020: Applied Mechanics

In the recent times, **engineering modelling** based on modelling and computer simulations has become indispensable tool for solving numerous problems, both of scientific and of engineering type. The key element of the currently carried research is the area focusing on modelling complex systems and physical phenomena as well as data acquisition and analysis. It is an area of versatile and interdisciplinary research trends with one of the mainstreams focusing on **applied mechanics**. It should be stressed that while modelling and next solving numerical models which describe complex phenomena of physical systems of various physical nature, **advanced calculation systems** constitute indispensable factor which allows obtaining correct solution.

The papers published in this book are a set of works presented at the 59th Symposium Modelling in Mechanics, which is one of the oldest recurring Polish conferences dealing with mechanical engineering. The need of annual meetings results, undoubtedly, from the vitality of mechanics as science about rules of motion and rest, and as the discipline which has great influence on human activity, life and the environment—eppur si mouve.

The greatest percentage of papers published in this book (nearly 35%) are academic works devoted to conducting experiments in mechanics. The most important purpose of computer simulation is prediction of physical phenomena or behaviour of technical systems. These predictions form the base of decision-making; however, the issues connected with verification and validation of numerical models through adequate experiments are of essential importance for their usefulness. The next group of about 30% of paper is formed by the ones which deal with expert systems based on, e.g. neuron networks or relational databases, elaborated for finding optimal solutions for a given issue and thus facilitating the process of taking rational decisions. The remaining papers are in majority on mechatronic systems and the use of finite elements methods in modelling mechanical systems. Currently, mechatronics is one of the most developing branches of contemporary science and technology.

The authors in their papers concentrate on the use of computer-aided design and virtual modelling, numeric simulations or methods of fast prototyping of mechanical structures of mechatronic systems, using at the same time the effect of synergy of research methods in order to get optimal geometric features and assure needed operation and maintenance parameters.

Acknowledgments. We wish to express our gratitude to Mariusz Pawlak for his professional and kind help with the preparation of the book.

December 2020

Contents

Mechatronics Systems Modelling Challenges and Threats

Paweł Bachorz$^{(\boxtimes)}$ (iD)

Department of Theoretical and Applied Mechanics, Silesian University of Technology,
ul. Konarskiego 18a, 44-100 Gliwice, Poland
Pawel.Bachorz@polsl.pl

Abstract. This paper includes the discussion on the modelling of mechatronic systems and introduces the term related to mechatronic systems. Following a brief introduction the next chapter analyzes some challenges and threats regarding modelling of such systems.

Keywords: Mechatronics system · Modelling

1 Mechatronics Systems

During development of civilization a human made use of simple machinery like a lever or windlass to make work simple. Over time the man constructed more complex mechanisms and devices, initially human and animal-powered devices or powered with renewable energy e.g. wind or moving water. However the dynamic development followed due to invention of steam–powered engine that triggered the industrial revolution. Roughly at the same time the first attempts were made to provide for engine speed automatic control with devices such as centrifugal controllers. Simultaneously development followed of mechanic systems providing transmission of energy from the engine to the working component.

The next phase was development of machines and devices driven with electric engines that featured with steady performance. Actually the engine itself provides for an electrical and mechanical system including a spinning rotor of specific mass corresponding to mechanical components, as well as stator and rotor winding corresponding to electrical system. Consequently as a result of development of electronic components, the first scalar control systems were designed, including the most known U/f method.

In this way this we approach the decade of the 80-ties of the last century, when a term of mechatronics was originally coined in Japan by T. Mori, an engineer of Yasakawa Electric Corporation and used initially in relation to electrical and mechanical system:

"The word, mechatronics is composed of mecha from mechanics and tronics from electronics. In other words, technologies and developed products will be incorporating electronics more and more into mechanisms, intimately and organically, and making it impossible to tell where one ends and the other begins" [9].

A. Mężyk et al. (Eds.): SMWM 2020, AISC 1336, pp. 1–10, 2021.
https://doi.org/10.1007/978-3-030-68455-6_1

In the age of computers and electronic devices, this term underwent an evolution what also resulted in adoption of new name related to young science. At present this term have the same meaning described with one simple 20-years old definition:

"Synergistic integration of mechanical engineering with electronics and intelligent computer control in the design and manufacturing of industrial products and processes" [5].

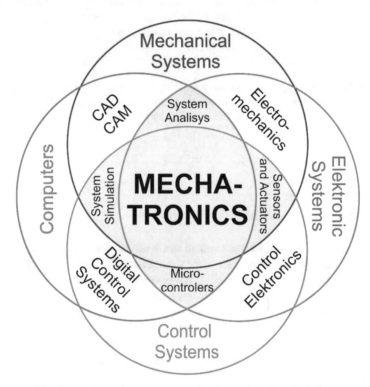

Fig. 1. Mechatronics as synergy of some engineering disciplines.

At present mechatronics is a multidisciplinary branch of engineering that focuses on the engineering of both electrical and mechanical systems and includes also a combination of other disciplines (see Fig. 1). We did not list all disciplines as it is impossible to name all of them. One can pay attention to the connection of mechatronics with humans in order to improve their quality of life. We can distinguish here the use of mechatronic devices in rehabilitation as devices supporting the rehabilitator or even replacing it, as well as in order to regain lost motor functions of the body, e.g. by using mechatronic prostheses [3, 6, 12].

Note also that most definitions used at present relate also to synergy. Synergy stands for the interaction or cooperation of some factors to produce a combined effect greater than the sum of their separate effects. A simple example may be the use of a vector control system in the operation of an AC motor, which allows for a significant improvement in the dynamics of the motor, i.e. reducing the amplitude of the force, faster reaching the target force value (Fig. 2 presenting a phase of engine start-up) [2].

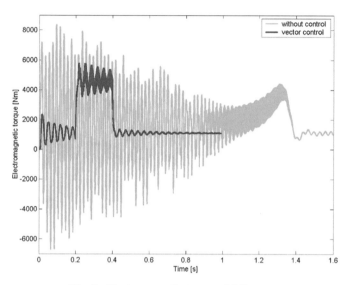

Fig. 2. Electromagnetic torque of AC motor.

Figure 3 and Fig. 4 describes some reference methods for mechatronic system simulation. Variant 1 presented the output system, variant 2 - focused on the mechanical part of the system, variant 3 - included the selection of vector controller settings, variant 4 is similar to mechatronic design - the mechanical part and the settings of the control system regulators were simultaneously optimized [8].

In each approach, it was possible to notice a reduction in the dynamic impact (amplitudes of forces and moments) in relation to the initial variant, which improves the durability and reliability of the device. However, option 4, representing the mechatronic approach, is interesting. By using the simultaneous optimization of the mechanical part and the settings of the control system regulators, we can clearly observe the resulting synergy (best visible on Fig. 4).

When describing mechatronics we focused on machinery, devices or created products. However this term covers much more for example production and designing processes. In the past designing of products included some works delivered in sequence that related to mechanical components, electric system, electronic components and finally control system. That approach required very often re-performance of some works and re-designing of the subassembly, as well as building many prototypes during verification of the whole system operation. At present works delivered during the initial design phase include the whole product, namely mechanic parts coupled with electronic components, as well as software and control system. Such product is designed, verified and optimized including all parts and systems together what eliminates re-performance of some works

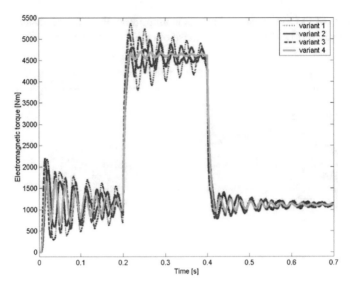

Fig. 3. Electromagnetic torque of AC motor with the vector control unit.

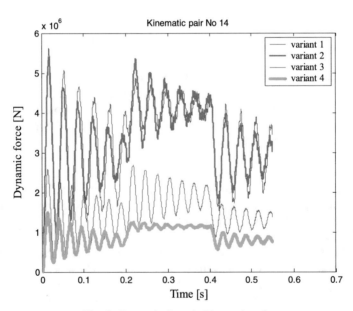

Fig. 4. Dynamic force in kinematic pair.

related to initial design phase, building of some prototypes and at the same time ensures synergy related effects. However such results cannot be obtained without works related to mechatronic system modelling.

2 Mechatronics Systems Modelling [1]

Mechatronic systems modelling is an inseparable part of designing or analysis of such systems. This phase may be replaced with experimental studies, which are sometimes costly and impossible. Sometimes there is no an appropriate object of studies, and sometimes tests would require putting the object out of operation what may generate very high costs related to downtime. Because the final effects of studies are unknown from time to time a non-steady state of the object of studies may result in damage or destruction of costly devices. In such cases modelling is essential. Moreover modelling prevents some costs related to studies performed with the object in scale. As we know it is not always easy and possible to apply obtained study results to the object under real conditions. It is evident in case of cube. Increasing twice the length of its sides will increase its volume eight times, and mass similarly, but e.g. side area will increase four times only.

In the modeling process based on the research object, we first prepare a physical model, then a mathematical model, finally implementing the mathematical model in the selected engineering software.

The physical model presents a part of the modeled system in a simplified way. The simplification allows to present the properties and structure of the tested system. The difficulty is the selection of the simplification. It is known that the influence of some of the variables describing the model is greater than others. Therefore, we can omit some of the variables without significantly affecting the functionality of the model. When dividing the analyzed system into subsystems, one should remember about the interactions and couplings between them. Preparation of a physical model requires the researcher to have thorough knowledge of the analyzed phenomenon or device.

Having prepared the physical model, we can proceed to the preparation of a mathematical model describing the processes, relationships and properties presented in the physical model using mathematical relationships.

Mechatronic systems ensure continuous control of speed and position of driver elements, what reduces dynamic load generated during gradual speed adjustment provided with traditional systems. Issues related to selection of structural features of mechatronic systems make for the biggest challenge during the design-construction process. The fundamental problems that follow during operation of complex systems relate to excessive vibration and consequently dynamic load caused by kinematic pairs, poor durability and reliability of the whole device. Therefore the crucial task of a design engineer is to select appropriate structural features to ensure desired dynamic characteristics of the systems. These problems may be solved with mathematical process models and optimization based on carefully determined criteria. The elaborated model of the systems shall represent the structure, at least a part undergoing optimizing, and moreover must include coupling of mechanic parts, electric components and control systems.

Modelling of mechanical and electromechanical systems as well as mechatronic systems underwent evolution similar to evolution of the term meaning mechatronics. Before we discuss mathematical process models, firstly we will focus on physical models. At the beginning these models related to a mechanical component, mostly one including a drive transmission which until today is very often provided with a mechanical gear. Initially linear models were developed that simplified reality significantly. These models offered a specific advantage namely simplicity enabling computation with a paper sheet only. Over time models also represented nonlinear features including variable gear mesh stiffness and backlash. As the drive system is not a mechanism suspended in the air, we need to consider also support provided with ball bearings embedded in housing or covers that make for other dynamic systems which may generate or reduce vibration– what actually most often is a disadvantage [1, 4, 11].

The main factor affecting precision of results and efficiency of dynamic studies is selection of the model representing the system. Toothed gears make for complicated dynamic systems featuring with many degrees of freedom. Consequently the detailed analysis of dynamic forces requires development of extended, nonlinear physical models. Note that the main problem faced during analysis at the initial phase of modelling relates to determining the parameters of the applied model that should represent exclusively the features of the real object that have essential impact on tested dynamic forces. Therefore depending on the purpose of studies and possibilities regarding identification of parameters, an adequate degree of simplification is applied. The main assumptions made for the purpose of simplification during modelling include:

- kind of external loads on the system,
- number of degrees of freedom and methods of system discretization,
- modeling the couplings between individual subsystems,
- physical relationships in kinematic pairs.

After elaboration of the physical model it can be expressed by equations to obtain a mathematical process model. Such model undergoes further simplification to eliminate some equations and ensure fast computation and progress of works, what unfortunately has a negative impact on precision of results. Actually simplification of the mathematical process model that minimizes a negative impact on results is a real art.

The above considerations focused on mechanical components of mechatronic systems. However it is quite similar in case of electrical and mechanical systems or electronic components. For example during modelling of electrical engines the initial models represented the linear characteristics of magnetic circuit only, whereas the next models represented also nonlinear performance of magnetic circuit. At present engine polyharmonic vibration models are being developed, if only possible.

It is more difficult in case of digital control system modelling. Due to switches and specific performance of electronics it is very difficult to represent operation of these systems with some equations. However Symulink dedicated engineering software comes to our aid to ensure easy implementation and integration of control system.

Integration of all parts represented with the model plays a crucial role considering the purpose of modelling that is simulation of the mechatronic system operation often as early as at the design and construction phase to provide detailed data regarding whole system operation and performance before building a prototype. Another advantage is optimization and development of system along with all possible improvements and modifications.

The most important phase of optimizing is determining criterion of optimizing for the specific limits and selection of the set of decision-making variables. Depending on the manner regarding optimizing the most adequate method is selected and the relevant algorithm providing the solution of problem. In spite of many IT advancements optimizing of dynamic characteristics provided with structural models are still time con-suming, therefore precision of elaborated model, the method of optimizing and number of decision-making variables require compromise between precise description of studied phenomena and costs of works related to computation and obtained results.

At the end note also the above mentioned compromise between precision of the model and time or costs related to use thereof. The more extended model the more precise results were obtained, but time required for computation was longer and higher were costs related to required equipment. Another issue related to acquisition of data necessary for putting the model into operation, and impossibility to gather some data what prevents development of the most desired model.

3   Challenges and Threats

At this point it is good to consider some challenges and threats possible during modelling, simulation and optimizing of mechatronic systems.

One challenge mentioned earlier is selection of the model representing the unit. It is being said that modelling makes for a kind of art. As we already said the more extended and precise model the more accurate results are obtained corresponding to real conditions. However nothing is for free. The more extended model, the more input data is required. And as we know it is difficult to obtain some input data. And because acquisition of data poses some difficulties, sometimes dismantling of assembly or device is required to obtain data. In case of extended models more powerful and at the same time costly supercomputing equipment is required for the purpose of analysis or during simulation of operation. There is also the question regarding an analysis and interpre-tation of results – and maybe simpler models would work best at the beginning. The next issue relates to time required for computation and simulation. Extended models make for many equations that require solving and experience shows that time required for solving equations increases linearly in relation to number of equations. Actually this can be represented with an exponential function. We know that at present time makes for very precious asset, but these issues are widely known. And usually we are well aware of compromise between the precision of the model and time or its costs.

Dynamic development of computers and computing power contributed to elimination of some numeric errors. Several years ago it was a common practice to verify if a calculation error is not over the range of obtained results. Other issue is selection of computation method to the performed task. Note that computers perform all received commands, also ridiculous ones, but it is operators or humans who take the whole liability for computations.

In 2003 electronic voting was held in Belgium during election of federal authorities. And one unknown female candidate obtained a lot of votes in one of the towns. Actually it was nothing extraordinary, but finally it turned out that a number of votes outnumbered all persons entitled to vote. The investigation revealed that the number of illegitimate votes was 4096, namely 2^{12} and the reason identified was as a "random bit-flip in position 13 item in computer memory" [7, 10].

But what has this to do with modelling? Actually it was all about cosmic radiation that is a well-known problem faced by scientists involved in the study of space, and very often ignored by people, as magnetic radiation of the Sun and Earth atmosphere protects us against effects of cosmic radiation. But in fact this protection is not complete and we are exposed to some high energy particles. Miniaturization of integrated circuits caused less and less variance between energy state interpreted 0 or 1, and collision with high energy particle may result in change of energy state from 0 to 1 and vice versa. Cases have been reported where cosmic radiation caused damage of integrated circuit or the whole subassembly. Of course we may say that we are not affected by it and it should not concern us. But can we be sure of that? Have you never experienced some problems related to computer freezing or uncontrolled reboot? How can we be sure that it was not caused by cosmic radiation [7]?

You can minimize the problem in simple way with a specific type of computer data storage that can detect and correct the most-common kinds of internal data corruption named error correction code memory (ECC). Unfortunately this solution is not available in case of low cost computers. What can we do then? In fact personal computers are inevitable for modelling and simulation when solving simple problems and it will be also in future mainly due to related costs. However we should be aware of risk caused by computation related errors. When we do not know a solution to the problem we can consider computation redundancy and comparative analysis thereof. Actually error detection during one time simulation or computation process is beyond our possibilities, in particular in case of stochastic methods employed for the purpose of computation.

Can bit change be related only to a risk? Not necessarily. Everybody experienced it at last once in his life that he suddenly found in his head the simple solution for a very complicated problem bothering him for a long time Actually in case of bit change a solution may also be very simple, unique, unrepeatable and at hand, although we may not know neither time nor type of computation error.

We already mentioned that sometimes it is not easy to develop a model due to difficult acquisition of parameters, nature of operation or poor knowledge of the specific phenomena. This problem may be solved with Hardware-in-the-Loop technique which replaces the missing part with the real object coupled with the model where communication with computer is provided with I/O cards. In many cases fast communication cards are required namely real-time communication cards [3, 6].

Unfortunately these solutions are costly. But we must not make savings in this area. A good solution for example is a vector controller that anticipates drive status on basis of data from sensors. Note that because of communication delays any status considered future one may actually relate to the past.

This paper should also consider an area related to artificial intelligence. There are many research under way focused on use of an artificial intelligence for mechatronic systems optimization and control purposes. Moreover making use world computer network and its contents e.g. neural networks for the purpose of practice is very promising. These resources contain a lot of data that can be used for education. However what is considered an indisputable advantage of the Internet (availability, accessibility) provides for the big disadvantage on the other hand. We all know that it takes many efforts and time to find reliable source of information when we are inundated with huge amount of information. Very good example is Tay an artificial intelligence chatter bot, released to replay to other Twitter users on basis of its interactions with people. However following several hours this service was shut down because it posted offensive tweets (e.g. racist content). We all know the Internet has made information tremendously more accessible, but unfortunately finding useful information may be difficult sometimes.

4 Summary

To sum up we may say that modelling makes for a kind of art. And it is difficult to reach compromise between precision, time and costs of use. This paper mentioned some risks and expectations towards mechatronic system modelling. Unfortunately due to limited scope of this work we did not discuss that subject it detail. Certainly there are many other issues to be discussed and it is evident that small challenges often contribute to the change of our approach towards modelling.

References

1. Bachorz, P., Gąsiorek, D., Kawlewski, K., Mężyk, A., Rak, Z., Świtoński, E., Trawiński, T.: Minimalizacja drgań mechatronicznych napędów maszyn roboczych. (Minimization of vibrations in mechatronic drives of working machines). Wydawnictwa Naukowe Instytutu Technologii Eksploatacji - Państwowego Instytutu Badawczego, Radom (2012). (Biblioteka Problemów Eksploatacji)
2. Bachorz, P., Świtoński, E., Mężyk, A.: Reduction of vibrations in electromechanical drive systems with vector control. Transport (Vilnius) 19(6), 276–279 (2004)
3. Duda, S., Gembalczyk, G.: Concept of coupling the rehabilitation treadmill with foot pressure sensors. In: Świder, J., Kciuk, S., Trojnacki, M. (eds.) Mechatronics 2017 - Ideas for Industrial Applications. MECHATRONICS 2017. Advances in Intelligent Systems and Computing, vol. 934, pp. 111–118. Springer (2019). https://doi.org/10.1007/978-3-030-15857-6_12
4. Ferfecki, P., Zaoral, F., Zapoměl, J.: Using floquet theory in the procedure for investigation of the motion stability of a rotor system exhibiting parametric and self-excited vibration. J. Mech. Eng. 69(3), 33–42 (2019). https://doi.org/10.2478/scjme-2019-0027
5. Harshama, F., Tomizuka, M., Fukuda, T.: Mechatronics-what is it, why, and how? - and editorial. IEEE/ASME Trans. Mechatron. 1(1), 1–4 (1996)
6. Konopelska, A., Jureczko, M.: Application of surface electromyographic signals for electric rotor control. In: Świder, J., Kciuk, S., Trojnacki, M. (eds.) Mechatronics 2017 - Ideas for Industrial Applications. MECHATRONICS 2017. Advances in Intelligent Systems and Computing, vol. 934, pp. 248–257. Springer (2019). https://doi.org/10.1007/978-3-030-15857-6_25
7. Łomnicka, N., Stradowski J.: Kosmos psuje komputery? (Space destroys computers?), Focus Nr 6-7/298, pp. 38–41 (2020)
8. Mężyk, A., Bachorz, P., Świtoński, E.: Dynamic analysis of mechatronic drive system with asynchronous motor. Eng. Mech. 12(3), 215–221 (2005)
9. Mori, T.: "Mechatronics" Yasakawa Internal Trademark Application Memo, 21.131.01, 12 July (1969)
10. Rapport concernant les élections du 18 mai 2003. Sénat et Chambre des représentants de Belgique (2003)

11. Šlesar, P., Jančo, R.: Strength and dynamic assessment of cage and bearing for railway carriage. Strojnícky časopis J. Mech. Eng. **70**(1), 127–134 (2020). https://doi.org/10.2478/scjme-2020-0012
12. Turek, J., Daniszewski, M., Wolnicki, P., Machoczek, T., Jureczko, P.: Modelling the anthropomorphic mechanical hand. In: Świder, J., Kciuk, S., Trojnacki, M. (eds.) Mechatronics 2017 - Ideas for Industrial Applications. MECHATRONICS 2017. Advances in Intelligent Systems and Computing, vol. 934, pp. 427–434. Springer (2019). https://doi.org/10.1007/978-3-030-15857-6_42

The Effects of the Aeration Phenomenon on the Performance of Hydraulic Shock Absorbers

Piotr Czop[1] , Grzegorz Wszołek[2], and Mariusz Piotr Hetmańczyk[3]([⊠])

[1] AGH University of Science and Technology, Mickiewicza 30D-1 Street, 30-059 Cracow,
Poland
[2] WSB University, Cieplaka 1c Street, 41-300 Dabrowa Gornicza, Poland
[3] The Silesian University of Technology, Konarskiego 18A Street, 44-100 Gliwice, Poland
`mariusz.hetmanczyk@polsl.pl`

Abstract. This work introduces a method, based on the DOE experiment, for
assessing the risk of negative aeration by obtaining an experimental measure,
which was further correlated with a theoretical gas concentration measure intended
for mono- or twin-tube hydraulic shock absorber models. The presented results
allow a quantitative assessment method to be defined in order to understand and
reduce the aeration effects. The methodology of accurately measuring the effects
of the aeration phenomenon in order to validate a simulation model of a twin-tube
shock absorber is the focal point of this paper. The paper contributes to analytical
tools that allow the risk of aeration to be reduced and the shock absorber design
to be optimized against this negative effect. Moreover, the correlated theoretical
aeration measure, namely the gas concentration parameter, can be used at early
design and configuration stages when the shock absolver valve characteristics are
adjusted to achieve the required damping force. The paper describes the physics
of the aeration phenomenon, experimental set-up and definition of a measurement
system in both hardware and software aspects. The approach presented is based
on novel algorithmic post-processing of signals to compute the values of a scalar
measure for the phenomenon's qualitative features of interest, using data registered
on a dedicated test bench under the controlled variation of device parameters.

Keywords: Aeration · Shock absorber · Measurement system · Signal
processing

1 Introduction

Gas is included in shock absorbers, separately from the oil, to provide compressibility to
allow for the oil forced out by the rod displacement volume when the rod moves inside
the shock absorber during a compression stroke. Shock absorbers are typically filled
with nitrogen to reduce corrosion, oil oxidation and aeration. Nitrogen in new damper
units is pumped under pressure that depends on the rod diameter and typically amounts
to 3 to 4.6 [bar] relative pressure.

A. Mężyk et al. (Eds.): SMWM 2020, AISC 1336, pp. 11–30, 2021.
https://doi.org/10.1007/978-3-030-68455-6_2

This work focuses on the effects caused by the presence of gas bubbles in the hydraulic oil on the behavior of the force signal generated by the movement of the rod in a shock absorber. Shock absorbers are devices that dissipate mechanical kinetic energy of the rod motion through viscous friction of the incompressible working medium onto the solid structure surfaces of the device as radiated heat. If a soluble gas is present in the oil chamber under pressure higher than the normal atmospheric pressure ($p_0 = 1e^5$ [Pa]), the characteristic of the device deteriorates, at least temporarily, in a manner that so far has yet to be measured reliably and accurately. The main goal of this paper is to present and model the observed effect of the aeration phenomenon in shock absorbers.

The main goal is achieved by verifying that the aeration measure is applicable to the effect and that the same effect is visible in the simulation result. The presentation of the results of research on the aeration phenomenon is divided into major parts, which are: describing the measurement system employed, including the test bench construction, design of experiments along with the device preparation procedure and signal choice; discussing theoretical aspects of modeling, signal processing and data analysis related to the topic; as well as experimentally showing an effect so far observed only in simulation results, aeration in monotube-type shock absorbers.

One of the most important negative contributors to the low- and high-frequency ride performance of hydraulic shock absorbers used in vehicles is the combined aeration and cavitation effect [1, 2]. The delay in the build-up of damping force (pressure in the chambers) and the hysteresis loop in the force-velocity response are attributable to fluid compressibility, which is caused by the existence of either gas (aeration) or liquid vapor phase (cavitation) at certain stages of the stroking cycle. The presence of the trapped air or liquid vapor results in a large piston displacement before a significant pressure drop across the piston. Aeration and cavitation are complex phenomena which depend on a few factors, e.g. the purity of the liquid and the rate at which the liquid is stressed.

The low-frequency effect is the delay in the build-up of damping force or, equivalently, slower than expected increase of pressure in the chambers, and the hysteresis loop in the force-velocity response (Fig. 1b) attributable to abnormal fluid compressibility, which is caused by the existence of either gas (aeration) or liquid vapor phase (cavitation) at certain stages of the stroking cycle. The high-frequency effect, manifesting itself as the presence of excessive vibrations (Fig. 1c) and the emission of noise [3] is caused by an abrupt and catastrophic collapse of cavities trapped in the hydraulic liquid, and is attributable to the aeration and cavitation phenomena as well [4, 5]. The reference shock absorber model was excited using a sine-wave signal with a stroke of 80 [mm] and velocity of 0.3 [m/s]. The aeration is quantified by means of a free gas concentration in the oils and is represented by the χ parameter, which is defined in the paper by Eq. 24. If gas is dissolved in the oil and free gas is isolated in the separate hydraulic accumulator chamber, the parameter has a low valve, e.g. χ-1E-06 or less. When a twin-tube shock absorber is stroked for a few minutes and the oil temperature increases, then the free gas from the accumulator (reserve tube) penetrates into the oil and free gas simultaneously releases from the oil due to increased temperature and pressure (governed by the Bunsen solubility coefficient, Eq. 6). In the case of a mono-tube shock absorber, the gas can only release from the oil since the accumulator is separated from the pressure tube by a floating piston.

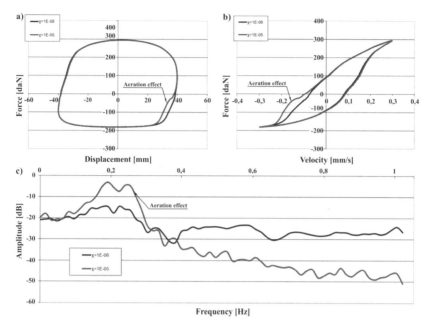

Fig. 1. Dynamic shock absorber characterization using: a) damping-displacement characteristic, b) damping-velocity characteristic, c) power spectrum vibration characteristic obtained for gas concentration χ parameter.

Shock absorber manufacturers use different data-driven methods to quantify the low-frequency aeration effect. There are essentially three commonly used measures: the length of the displacement after which the damping force achieves the expected values (point of return to the reference curve, Ford standard); the loss of damping energy between the first and the aeration affected measured force-displacement loop (Mazda standard); the length of the displacement from the maximal displacement to the point of change in the sign of the second derivative of the force signal (inflection point approach used in this paper, GM standard).

It is also possible to use the Ford specification to define a variant of an energetic measure or to aggregate (consider a weighted average) various measures.

In the paper, in order to establish the necessary theoretical background, elements of the gas solubility theory, as well as the theory of compressible, sub-sonic fluid flow applicable to operation principles for a shock absorber, are recalled. The mathematical description shown in Sect. 2.2 clearly indicates factors influencing the phenomenon, thus indicating the direction in which the measurement system is to be developed and validated. The means by which the purpose of this paper has been achieved is a combination of the principles of signal processing, statistics and hydraulics. The properties of fluid and fluid-bubble emulsion flowing through restrictions are derived based on [2]. Shape identification algorithms using smoothing FIR filters are also described in this work, while all results are presented using a statistical design of experiment point of view [6]. State-of-the-art techniques are employed to resolve the key difficulty of the topic at hand, which is to quantitatively assess the magnitude of the aeration effect; the difficulty that lies in the necessity to quantify an intrinsically qualitative effect.

2   Theoretical Background

2.1   Shock Absorber Device

A shock absorber is a device that dissipates mechanical kinetic energy of linear motion through viscous friction of the working medium onto the device's solid structure surface in the form of heat dissipated from its exterior surfaces [2]. A shock absorber, in its generic form, is an assembly of a piston moving inside a liquid-filled cylinder (Fig. 2).

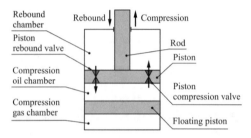

Fig. 2. Hydraulic damper scheme.

As the piston is forced to move inside the cylinder (the pressure tube), a pressure differential is built across valve assemblies, forcing the liquid to flow through restrictions (orifices) and valves. The piston divides the volume of the pressure tube into two chambers: the rebound chamber, which is the part above the piston, and the compression chamber, the part below the piston. Since the liquid is essentially incompressible, the volume of the rod entering the liquid filled cylinder has to be compensated by compression of the gas located in the interior of the device, in the volume called the gas chamber. Depending on the design, the gas may be separated from the oil by an air-tight separator, or have direct contact with the oil.

The research device utilized in this work, namely an equivalent shock absorber physical model, has been equipped with an air-tight gas chamber separator, yet the gas valve installed additionally in the base of the device allows additional gas to be applied to the oil chamber in a controlled way (see Fig. 4b). Thanks to the introduction of such a modification, a contact surface between gas and oil is ensured and the amount of gas dissolved is controlled.

2.2   Gas Solubility

The term most commonly used to name the phenomenon which is the focal point of this paper, "aeration", is somewhat misleading, as it refers to the cause and not the effect of the investigated phenomenon. As a matter of fact, the phenomenon results from releasing pressurizing gas dissolved in the working medium in the form of gas bubbles to form an emulsion [7]. Such an emulsion is not homogeneous and has a certain life-span during which, at least partial, re-adsorption of bubbles into the working liquid occurs. Bubble formation and bubble re-adsorption is caused by local changes in the pressure in shock absorber chambers with respect to the pressure-at-rest, as well as by the flow through restrictions, also relative to the resting state. In this paper, we refer to the term aeration in a broader sense, namely we refer to all the phenomena related to gas and oil mixing, gas dissolution, bubble formation, bubble re-adsorption and flow of the bubbly oil through internal chambers and restrictions as aeration.

In a reservoir chamber of a typical shock absorber, the hydraulic oil is in direct contact with gas, i.e. there is an interfacing surface between the working medium (normally hydraulic oil) and the pressurizing medium (gas). Gas is introduced to the interior under certain pressure and, in the resting state, is separate from the oil. The gas in the reservoir chamber, which is a compressible fluid, allows compensation of the volume of the rod entering the interior of the shock absorber during its work. Additionally, the balance between valves used in typical shock absorbers depends on the gas pressure in the accumulator.

A liquid that was exposed to a soluble gas, i.e. a liquid that for a period of time had a contact surface with the atmosphere of a gas that can dissolve in it, can be in one of the following three forms, depending on the instantaneous pressure conditions. Those are a liquid-gas solution, a liquid-gas bubble emulsion and foam.

The liquid-gas solution is prone to bubble formation when the local pressure of the liquid-gas solution decreases below the so-called saturation pressure. One should notice that bubble formation is a spatiotemporal phenomenon, in the sense that the pressure in the flow of a fluid is a function of all spatial variables and time, i.e. $p = p(x, y, z, t)$, and is, in general, governed by the notorious Navier - Stokes equation (NSE) in its compressible form, which takes the form of

$$\rho \left(\frac{\partial v}{\partial t} + v \cdot \nabla v \right) = -\nabla p + \mu \nabla^2 v + \left(\frac{1}{3} \mu + \mu^v \right) \nabla (\nabla \cdot v) + f \tag{1}$$

where ρ is the density, v is the velocity field, p is the pressure field, μ is compressibility, f represents external forces and v is the viscosity.

Moreover, since the solubility of a gas in a liquid is directly proportional to the absolute pressure above the liquid surface (Henry's Law), and normally decreases with rising temperature [2] all of the mentioned liquid-gas mixtures, when put in motion, can be considered as liquid with pockets of gas. Liquid with gas bubbles can no longer be treated as effectively incompressible, so the presence of gas bubbles is the cause of the damping force loss in the shock absorber. It is an undesirable and negative effect visible as an asymmetry of the force to displacement curve and, as such, should be minimized during the design process. Figure 3 illustrates the influence of the oil aeration effect on the shock absorber performance based on the force-displacement characteristic (the results of the simulation were obtained for $\chi = 2.95 \cdot 10^{-4}$).

The aeration effect causes a drop in the damping force visible in the top-left corner of the characteristic (Fig. 2). Using the Ideal Gas Law

$$p \cdot V = n \cdot R \cdot T = \frac{m}{\mu} \cdot R \cdot T \tag{2}$$

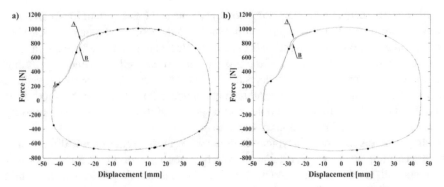

Fig. 3. A force displacement curve (the A line) with its inflection points (black dots) computed based on the Savitzky-Golay smoothing displayed for the reference (the B line): a) the result for the experiment <Run 2> (cf. Table 1) in the DOE matrix, b) the corresponding result obtained from the simulation model.

and Henry's law

$$m_{max} = c_{GA} \cdot V_{oil} \cdot p_{acm} \qquad (3)$$

one obtains the relation between the injector pressure and volume and the tooling parameters that ensures that the maximal amount of gas is dissolved in the oil

$$p_{inj} \cdot V_{inj} = \frac{c_{GA} \cdot V_{oil} \cdot (p_{acm} - 1)}{\mu} \cdot R \cdot T = \frac{c_{GA} \cdot R \cdot T}{\mu} \cdot V_{oil} \cdot (p_{acm} - 1) \qquad (4)$$

Dimensional analysis indicates that

$$\frac{[c_{GA}] \cdot [R] \cdot [T]}{[\mu]} = \frac{kg/(m^3 Pa) \cdot J/(mol \cdot K) \cdot K}{kg} = \frac{J \cdot kg}{kg \cdot mol \cdot (m^3 \cdot Pa)} = \frac{N \cdot m}{mol \cdot (m^3 \cdot N/m^2)} = 1 \qquad (5)$$

For typical hydraulic oils, Henry's constant is of the order of unity [2]. The value of the constant in the equation above is sometimes called the Bunsen coefficient. It is not hard to calculate that, for typical oils, the Bunsen coefficient is

$$\alpha_V = \frac{c_{GA} \cdot R \cdot T}{\mu} \approx 0.08 \qquad (6)$$

where c_{GA} is the so-called Henry's constant, V_{oil} is the volume of the liquid phase in which the gas dissolves and p_{gas} is the pressure of gas above the liquid surface. The variables V_{inj}, V_{oil} are given parameters of the design and available tooling. One is to take the mass of gas already dissolved in the given volume of oil into account.

In detail, the mass of gas that can additionally dissolve by pressurizing the container is the mass of gas dissoluble minus the mass of gas already dissolved.

2.3 Automatic Shape Identification

The method described herein [8] assumes that the geometrical features of a differentiable curve in a real, two dimensional space affected by an added noise of a broad spectrum are to be computed. This assumes that a curve γ is parameterized as

$$\gamma(t) = \big[x(t), y(t)\big] \in R^2, t \in I \subset R \qquad (7)$$

where R denotes the real line. The curve γ is of the class $C^r(I)$, i.e. it is a function continuously differentiable up to the order r, if and only if both functions $x = x(t)$ and $y = y(t)$ are of the class $C^r(I)$ over their domain $I \subset R$. In this paper, we are particularly interested in class $C^2(I)$ curves. Throughout the paper, we assume that the noise is added, i.e. that the measured curve is the sum of a smooth curve γ and a noise. We call the noise additive as we are going to assume that

$$\gamma_e(t) = \big[x(t) + e_1(t), y(t) + e_2(t)\big], t \in I \subset R \qquad (8)$$

where $e_i = e_i(t), i \in \{1, 2\}$ stand for noise signals, possibly different for both coordinates. The purpose of the algorithm discussed in this paper is to eliminate as much noise as is needed to find the 'dominating' shape of the curve, yet not to overly 'flatten' it. For the purpose of constructing the algorithm, consider the discretization

$$\gamma_i = \gamma(t_i) = \big[x(t_i), y(t_i)\big] \qquad (9)$$

of the curve γ evenly spaced in time. Here $t_i = t_0 + i \cdot \Delta t, i \in \{1, \ldots, N\} \subset N$ and $\Delta t = const$ is the time step.

The problem discussed in this work can be expressed in the language of differential geometry, as a search of convex/concave segmentation of the curve that is for looking for inflection points of a parametric curve [9]. Let's recollect the conditions for convexity of a curve defined parametrically as in Eq. (7). The necessary condition of existence of the inflection point is the existence of a solution to the equation

$$0 = \begin{vmatrix} x'(t) & y'(t) \\ x''(t) & y''(t) \end{vmatrix} = x'(t) \cdot y''(t) - x''(t) \cdot y'(t) = D(t) \qquad (10)$$

2.4 Savitzky-Golay Filters

In this section, a particular type of low-pass filter, well-adapted for data smoothing, and variously termed Savitzky-Golay (S-G), least-squares, or DISPO (Digital Smoothing Polynomial) filters is described. The class of S G filters is derived directly in the time domain, i.e. without use of the Fourier space techniques, by a generalization of the moving window averaging technique [10] Such a generalization has certain advantageous properties that allow the preservation of differential-geometric properties of smoothed signals.

In order to construct the filter, consider a series of data values equally spaced in time $f_i = f(t_i)$, where $t_i = t_0 + i \cdot \Delta t, i \in \{0, \ldots, N\} \subset N$, for some constant sample spacing Δt and an index i being a finite natural number. The simplest type of digital filter, the so-called non-recursive, or finite impulse response (FIR) filter, replaces each data value f_i by $\hat{f}_i$, being a linear combination over a set $G_i =$

$\{g_i = f_{i+k} \vee k = -K, \ldots, 0, \ldots, J \wedge K, J < N/2\}$ of some number of nearby neighbors. In that sense, constructing a digital filter means finding a set $C = \{c_k\} \subset R$ of coefficients in such a way that the filtered value

$$\hat{f}_i = \sum_{k=i-K}^{i+J} c_k \cdot f_k \tag{11}$$

satisfies a certain predefined set of criteria. In this paper, we will refer to the simplest case, i.e. the case where $c_k \equiv 1/(J + K + 1)$, called moving window average, as the linear FIR filter. The name is justified, as the coefficients result from finding the best fitting constant function. From a differential-geometric point of view, linear FIR filters introduce bias if the underlying function has a non-zero second derivative. It is not difficult to notice that, for instance at a local maximum, the moving window averaging always reduces the value of the underlying function. The moving window averaging does, however, preserve the area under a spectral line, which is its zero-moment, and if the window is also symmetric, i.e. if $J = K$, its mean position in time, which is its first moment. What is violated is the second moment, equivalent to the line width.

The idea of Savitzky-Golay filtering is to find the set $C = \{c_k\}$ of filter coefficients that preserve higher moments. Equivalently, the goal is to approximate the underlying function f within the moving window not by a constant (whose estimate is the average), but by a higher-order polynomial. For typical applications, where knowledge of the first and the second-order derivative is needed, a quadratic polynomial is sufficient. In the examples presented in this paper (Fig. 3b), the 3rd-order derivative is necessary and so a higher degree polynomial is needed. Details of the derivation of smoothing coefficients, as well as coefficients of derivatives, are outside the scope of this paper; interested readers are referred to [10].

2.5   Signal Processing

Iterative, or adaptive, use of the Savitzky-Golay smoothing filter lies at the core of the data processing method applied to measurement signals registered in the experiments underlying this paper. The method is capable of distinguishing the dominating shape of the noise curve. Properties of the S-G filter allow for very accurate recognition of the shape of a given curve even in the presence of significant noise, like high frequency or high amplitude variation in the signal occurring during valve opening. It is a well-known fact from differential geometry [9] that shape features are mostly characterized by inequalities on a differential form involving derivatives up to the second order. The major difficulty with finding the inflection points, thus determining the length of the concave behavior of the force signal, is the influence of measurement noise and high frequency oscillations, caused by valve opening and stabilization of the disc stack, on the inflection point selection method. Iterative use of the Savitzky-Golay filter permits selection of only the most relevant inflection point. Filter parameters (degree k of the smoothing polynomial and size n of the frame) were adjusted iteratively to obtain as low a number as possible of inflection points from which the free stroke was calculated, relative to the length of the test stroke. The distance between the minimal displacement and the concave-to-convex inflection point refers to the undamped movement.

2.6   Statistical Methods in the Design of Experiments (DOE)

An experiment is a deliberate change of some process variables [11], called factors, in order to observe the effect on some response variables. The statistical design of experiments (DOE) is an efficient procedure for planning experiments so that the data obtained can be analyzed statistically yielding objective conclusions; objective in the statistical sense, that is, with a confidence interval associated to its value [11]. Underlying this paper, a methodology of investigating the joint dependency of the measurement results on multiple parameters has been devised and validated. The DOE, as a methodology, considers all the variables simultaneously and is capable of predicting response over wide ranging regions in the parameter space. In comparison to other techniques, the DOE provides information about the interaction of factors and the way the total system works, something not obtainable through testing one factor at a time while holding other factors constant (OFAT).

The statistical theory underlying DOE is intrinsically connected with the concept of process models. A process model is a function of several variables, typically real-valued. Model input variables can be grouped into two categories, controlled inputs (factors) and uncontrolled noise. Typical linear models with two factors are expressed as

$$Y = f(X_1, X_2, \varepsilon) = \alpha_0 + \alpha_1 X_1 + \alpha_2 X_2 + \alpha_{12} X_1 X_2 + \varepsilon \tag{12}$$

where X_1, X_2 represent controllable factors and ε is the cumulative term representing the noise. A model of this type is used in the following sections of the paper to present the research results. The DOE is equipped with a center-point to allow the measurement data following a non-linear model instead to be checked. Identification of a non-linear model requires a different design matrix than that shown in Table 1 and is much more costly as it requires twice as many test runs to be performed.

3   Formulation of the Mathematical Model

The process of formulating a simulation model consists of three phases, namely: formulation of the mathematical representation of first-principle laws; model adjustment and calibration process; model validity assessment.

This section defines the mathematical formulation of a simulation model of a mono-tube shock absorber device in which a two-phase flow of the emulsion of gas bubbles in the hydraulic oil is foreseen. The following are assumptions under which the presented model of the shock absorber has been developed: dependency of density on pressure is non-linear (emulsion of gas bubbles in the oil); pressure and density are uniformly distributed in each particular chamber; pressure-flow characteristics of all restrictions are given as monotonic functions; valves open and close abruptly in a completely symmetrical manner (valve dynamic is not considered); oil temperature is constant; friction between the floating piston and pressure tube is negligible (very small compared to other friction forces ensured by installation of low friction sealing and the lack of presence of any side force); mass of the floating piston is negligible compared to the mass of the oil, so that inertia is also neglected.

3.1 Force Equilibrium Equation

The behavior of a mono-tube shock absorber with a top-mount attached to the rod is described by the equation [5]

$$m_1 \cdot \ddot{x}_{rod} + c_{TM} \cdot (\dot{x}_{rod} - \dot{x}_{TM}) + k_{TM} \cdot (x_{rod} - x_{TM}) = F_d \qquad (13)$$

where $m_1 = m_{rod} + 30m_{TM}$ [kg] is the mass of the rod assembly (the rod, the piston and the complete valve assembly) and the mass of that element of the top-mount that is affixed to and moves together with the rod assembly, c_{TM} [N•s/m] is the top-mount damping coefficient, k_{TM} [N/m] is the top-mount stiffness, x_{rod} [m] is the rod displacement, x_{TM} [m] is the top-mount displacement (displacement of the part of the top mount part fixed to the car or test machine) and F_d [N] is the force generated by the shock absorber. The top mount is an elastic element used to mount the shock absorber device to the remaining elements of the suspension system and provides better noise-reduction capabilities [5].

The force F_d generated by the shock absorber is obtained by taking the equilibrium of forces acting on the inertial piston-rod assembly, given by

$$F_d = p_{reb} \cdot A_{reb} + p_0 \cdot A_{rod} - p_{com} \cdot A_{com} + F_{fric} \qquad (14)$$

where $A_{rod}, A_{com}, A_{reb}$ [m2] represent areas of the rod, the compression and the rebound sides of the piston, respectively, and p_{com}, p_{reb} [Pa] pressures in the compression and the rebound chambers, respectively, while p_0 [Pa] is the atmospheric pressure. The quantity F_{fric} [N] stands for the sum of all dry friction forces between the piston and the pressure tube and between the rod and the rod guide. Since the total dry friction force F_{fric} [N] depends on the direction of relative rod travel, which is, however, independent of the velocity and piston position, it is modeled as

$$F_{fric} = F_{fric_{max}} \cdot tanh\left(\frac{\dot{x}_{tube} - \dot{x}_{rod}}{v_{ref}}\right) \qquad (15)$$

The maximal friction force $F_{fric_{max}}$ [N] is obtained experimentally using only the piston-rod assembly with no valves installed. The friction between the floating piston and the pressure tube is omitted because it is significantly smaller than F_{fric} [N].

3.2 Modeling Flow Through Restrictions

Characteristics of the compression and rebound valves of manufactured shock absorbers are normally tuned to ensure that customer specifications regarding force and durability are met. The force characteristic is given as a function of force versus velocity, while durability is typically specified in the number of sinusoidal cycles of a given amplitude which the shock absorber has to withstand. In this section, the experimental model of a valve in the form of static pressure-flow curves is utilized in flow modeling, even though a number of phenomenological parameters, such as discharge coefficients dependent on variable geometry, are required to be found. Readers interested in existing alternatives concerning static and dynamic models of valve systems can find them in [12]. Static pressure-flow characteristics of all restrictions need to be measured for the purpose of capturing the relationship between the pressure drop Δp across the considered valve assembly and the volumetric flow rate q through it.

Changes in the oil mass in the compression and rebound chambers are obtained using the following mass flow equation

$$\acute{m}_{reb} = \left(q_{piston,com} - q_{piston,reb} \right) \cdot \varrho_{reb,emu_{com}} \qquad (16)$$

where m, $\acute{m}$, q, ϱ represent the mass [kg], the mass flow rate [kg/s], the volumetric flow rate [m^3/s] and the density [kg/m^3] of the oil flowing through the valves, respectively. The density ϱ is obtained as the average value of the oil density in the connected chambers, which is justified by the fact that the maximal calculated change in the density of the oil is less than 0.35% if the differential pressure over the valve does not exceed 10^5 [Pa] [12]. Rearranging the mass flow equation, the following differential equations are obtained

$$\acute{m}_{reb} = \frac{1}{2}q_{piston} \cdot \frac{m_{reb}}{V_{reb}} + \frac{1}{2}q_{piston} \cdot \varrho_{emu_{com}} \qquad (17)$$

where $q_{piston} = q_{piston,com} - q_{piston,reb}$. The volume V_{reb} [m^3] depends on the geometry and its initial value is determined as

$$V_{reb} = V_{reb_{ini}} + A_{reb} \cdot (x_{rod} - x_{tube}) \qquad (18)$$

The average density of the oil-gas emulsion in the compression chamber $\varrho_{emu_{com}}$ [kg/m3] is a function of the average density of the oil-gas emulsion in the compression oil chamber and gas in the compression gas chamber ϱ_{com} [kg/m^3]. The average density ϱ_{com} can be calculated using the formula

$$\varrho_{com} = \frac{m_{gaz_{com}} + m_{emu_{com}}}{V_{com}} \qquad (19)$$

where $m_{emu_{com}} = m_{emu} - m_{reb}$ and $V_{com} = V_{com_{ini}} + A_{com} \cdot (x_{tube} - x_{rod})$ with $m_{gas_{com}}$ [kg] representing the mass of the free gas in the compression gas chamber, $m_{emu_{com}}$ [kg] representing the mass of the oil-gas emulsion in the compression oil chamber, m_{emu} [kg] the mass of the oil-gas emulsion inside the mono-tube shock absorber, m_{reb} [kg] the mass of the oil-gas emulsion in the rebound chamber and V_{com} [m^3] the volume of the compression chamber (sum of oil and gas chamber). The volumetric flow q [m^3/s] through the piston depends on the pressure drop Δp and is given as a static characteristic that is determined experimentally and the pressures p_{com}, p_{reb} are determined as functions of density.

3.3   Aeration Effect as a Two-Phase Flow

The model proposed in the paper takes into account only the aeration effect using a homogeneous oil-gas model. The homogeneous model assumes that the gas and liquid have the same velocity and are in the same thermal equilibrium. Additionally, the solubility of gas in liquid is constant and equal in all chambers. The solubility of the gas in the oil can be measured by the dimensionless χ value. The χ value is the ratio of the mass of gas to the total mass of oil and gas. Using an empirically calculated (based on Henry's Law), or obtained by experiment, value of χ, the density ϱ of a homogeneous gas-oil mixture can be calculated with the following formula

$$\varrho = \varrho_{emu} = \left(\frac{\chi}{\varrho_{gas}} + \frac{1 - \chi}{\varrho_{emoilu}} \right) \tag{20}$$

with $\varrho = \varrho_{reb} = \varrho_{emu_{reb}}$, $\varrho_{gas} = \varrho_{emu_{gas}}$ and $\varrho_{oil} = \varrho_{oil_{emu}}$.

Using the above expressions, the pressure-density dependency in the rebound chamber can be determined in a few steps. In the first step, the values of pressure are calculated for assumed oil densities using Eq. (8). Next, the density of gas dissolved in the oil is determined using the Ideal Gas Law [2]. Finally, Eqs. (22), (23) and (24) are solved for p_{reb} as a function of ϱ_{reb}.

Determination of pressure-density dependencies in the compression chamber requires additional steps. The total mass of gas (gas in the compression gas chamber and gas in the gas-oil emulsion) is obtained from the Ideal Gas Law. The mass of gas in the compression gas chamber is given by

$$m_{gas_{com}} = m_{gas} - \chi \cdot m_{oil} \tag{21}$$

For the expected range of oil density $\varrho_{oil_{com}}$, the pressure p_{com} in the compression chamber is given by

$$p_{com} = p_{ini} + K \ln \left(\frac{\varrho_{oil\_com}}{\varrho_{ini}} \right) \tag{22}$$

The density of the gas in the compression gas chamber and gas in the gas-oil emulsion in the compression oil chamber is determined from the Ideal Gas Law, and the average density of the oil-gas emulsion in the compression chamber is calculated based on the densities computed in the same way as (21) with $\varrho = \varrho_{com} = \varrho_{emu_{com}}$, $\varrho_{gas} = \varrho_{gas_{com}}$ and $\varrho_{oil} = \varrho_{oil_{com}}$.

The volume $V_{emu_{com}}$ and the mass $m_{emu_{com}}$ of the homogeneous oil-gas emulsion in the compression oil chamber are easily computed. Finally, the average density of the homogeneous oil-gas emulsion in the compression oil chamber and gas in the compression gas chamber is

$$\varrho_{com} = \left(\frac{\chi_{com}}{\varrho_{gas_{com}}} + \frac{1 - \chi_{com}}{\varrho_{emu_{com}}} \right)^{-1} \tag{23}$$

where

$$\chi_{com} = \frac{m_{gas_{com}}}{m_{emu_{com}} + m_{gas_{com}}} \tag{24}$$

The model is valid and capable of accurately simulating the aeration effect, as shown in Table 1, by showing a value of the gas concentration parameter χ corresponding to the value of the aeration measure obtained in the physical experiment. Because of the nature of the dependency of the aeration measure in the simulation model on the value of the gas concentration parameter χ, the authors decided to use a logarithm of the base 10 of the value. In each search for the value of χ corresponding to the measured aeration, all other parameters were set identical to parameters in the physical experiment. Figure 3b shows an illustrative example of the simulation result that corresponds to the measurement result displayed in the left pane of the figure. The result of the second run of the DOE experiment is presented (Table 1).

4 Experimental Setup

4.1 Tooling

For the purpose of validating the aeration model and the measurement data-processing algorithms, as well as to illustrate its applicability in design optimization, a special, dedicated modular tooling has been created (Fig. 4).

The tooling retains all the working principles of a mono-tube shock absorber, differing however substantially from any practically implementable device both geometrically and parametrically. The tooling is also capable of emulating effects observed in twin-tube shock absorbers. The tooling in its current shape allows much better control of design parameters than any production or research unit available on the market. Readers interested in the theory and engineering practice of hydraulic damping devices are referred to [2]. During the design process, assumptions concerning the ability to take pressure measurements in the interior of each of the hydraulic chambers have been foreseen. Pressure changes can be registered thanks to pressure transducers installed in the wall over and under the piston.

The original design of the tooling, the mono-tube design, has been adapted by introducing a gas injection valve located on the wall below the pneumatic accumulator. This was done in order to introduce sufficient mass of gas to the oil chamber to dissolve in the oil and to achieve a gas saturation state. In order to saturate the oil with soluble gas, the maximal soluble mass of the gas has to be introduced; the maximal soluble mass of the gas is given by the Henry's Law [2]. Having the ability to saturate the oil with the gas allows research to be conducted on the twin-tube type device.

Measurement was performed at a fixed and known temperature by filling the interior with oil completely and taking the total mass. The volume of oil was calculated based on the known temperature-density characteristics of the oil by subtracting the mass of the empty tooling from the measurement result. The accuracy of the scale used amounts to 1 [g] and is sufficiently accurate for the purpose.

4.2 Measurement Chain

A servo-hydraulic load frame equipped with force sensors, which is capable of controlling the actuator displacement, has been employed in order to measure characteristics of the dynamics of the especially designed shock absorber tooling described in the previous section. The actuator, manufactured by the MTS Company, type 858, allows any type of forcing signal to be executed and either a deterministic or random signal can be followed by the actuator. Control of the actuator movement is possible thanks to a

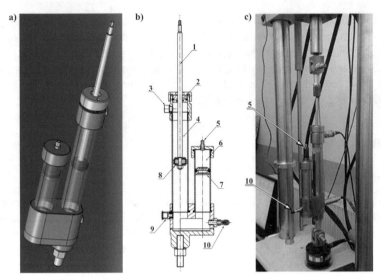

Fig. 4. Schematic display of the tooling employed in the research: a) the 3D design model, b) cross-section, c) overview of the device mounted on the test machine; where: 1 - rod, 2 - sealing, 3 - sensor hub, 4 - oil chamber, 5 - pneumatic valve, 6 - accumulator, 7 - separator piston, 8 - damper valve, 9 - sensor hub, 10 - air valve for gas implementation.

servo-valve of flow capacity up to 63 L/s under the nominal pressure of 21 [MPa]. The servo-hydraulic load frame is controlled electronically by a dedicated data acquisition system, called MTS Flex Test, working under the control of the "MTS Model 793" software, which collects data and controls values of the force or the displacement of the hydraulic actuator through the strain gauge-based load cell and an LVDT sensor.

The MTS 858 test bench is capable of reproducing the shape of the desired sinusoidal signal under the load of the test shock absorber up to the velocity of 0.7 [m/s] with sufficient accuracy. For velocities higher than that, the shape is distorted to some extent and maximal velocity is not always attained. For the purpose of this investigation, parameters of the test bench are sufficient, as sinusoidal signals of maximal velocity 0.5 [m/s] are used in the main part of the research. Two types of dynamic tests had to be performed; the preliminary test to obtain valve force-velocity characteristics and check valve stability, i.e. valve balance under a specified pressure of $p_{acm} = 12 \cdot 10^5$ [Pa] in the pneumatic accumulator, followed by the main test performed according to the DOE matrix. The preliminary test was carried out with pure oil, i.e. oil with no additional gas dissolved, by executing a sinusoidal signal of amplitude A = 45 [mm] in three sequences of 5 cycles each with maximal velocities of 0.5 [m/s], 0.7 [m/s] and 1 [m/s]. Sampling frequency was set to 1024 [Hz]. The DOE experiment used a sine signal of amplitude 45 [mm], maximal velocity 0.5 [m/s] carried for 100 cycles with the same sampling frequency of 1024 [Hz].

4.3 Design of the Experiment

Research work and testing for this paper were carried out according to a DOE matrix in which the following were the control parameters: the volume of the oil and the pressure of the gas in the accumulator. In order to validate theoretical considerations and simulation results, a series of experiments were performed in a manner organized according to the matrix shown in Table 1.

Table 1. Full factorial DOE matrix.

Run Order	Centre Point	V_{oil}[ccm]	Pressure in accumulator $p_{acm} \cdot 10^5$ [Pa]	Aeration Measure [%]	Gas concentration measure $log \chi$ [-]
1	yes	390	8	21.1	−4.7
2	no	410	11	29.4	−3.5
3	no	410	5	27.8	−3.3
4	no	370	11	19.0	−5.7
5	no	410	11	33.1	−3.4
6	no	370	11	19.3	−5.4
7	yes	390	8	21.0	−4.7
8	no	370	5	14.7	−5.5
9	no	410	5	27.0	−3.7
10	no	370	5	14.5	−5.8

A full factorial DOE matrix with two parameters and a center point, with two repetitions of each experiment and randomized runs were conducted. The penultimate column in Table 1 shows numerical values of the aeration measure obtained in the physical experiment, whilst the last column contains values of the natural logarithm of the value of the gas concentration parameter χ, which in the simulation model yields an aeration measure equal to the experimental one. The amount of gas added to the oil chamber was selected to equal the maximal level under the given gas pressure and oil volume conditions. In other words, experiments were carried out under the assumption that, in the initial state, the oil is saturated with the gas under the experiment conditions. In order to achieve saturation, a fixed volume of gas, under pressure prescribed by (4), was pumped into the hydraulic chamber and then the accumulator pressure was applied. The choice of parameter levels in the design is sufficient to distinguish the effect of magnitude 3.88 at the test power of 95%.

Experiments of the DOE were preceded by a series of valve optimization experiments in which a disc-spring combination was found that ensured stable performance of the valves, i.e. it yielded the expected force-velocity curve without abnormal hysteresis. The results of those experiments are not shown, as they are beyond the scope of the text. Table 1 shows, alongside the DOE design matrix, values of the aeration measure, which is the experiment response and the corresponding value $log_{10}\chi$ of the base-10 logarithm of the most important parameter of the simulation model, the ratio (24) of the mass of gas in bubbles to the total mass of the emulsion. The value of the parameter χ was found in such a way that the value of the aeration measure calculated from the simulation results equals the corresponding value in the physical experiment, with all other parameters in the model equal to respective parameters in the physical experiment. The degree to which simulation fits the measurement data in this case is quantified by the correlation coefficient which amounts to 94.5. Graphically, the fit quality is presented in Fig. 6, where the line is fitted to the data with $R^2 = 89.3$. Such a condition constitutes a necessary condition for model validity; the sufficient condition in this case is that, for a given value of the aeration measure in the physical experiment, there is only one value of the parameter χ for which the aeration measure obtained from the simulation model is equal. In the case shown herein, the model validity is confirmed - the data in Table 1 ensures the necessity, while the monotonicity of the dependence of the simulated aeration on χ ensures the sufficiency.

It is easy to notice in the DOE matrix that both parameters require assembly of the tooling before each experimental run. In order to do so, the following procedure was used to build a test sample. Firstly, oil temperature was read when the tooling was assembled and sealed. Secondly, the tooling was filled with the appropriate amount of oil according to Table 2 using the scale and known density-temperature relation.

Table 2. Values of the accumulator pressure, gas volume and injected pressure for a particular run.

Volume of oil V_{oil} [ccm]	Pressure in accumulator $p_{acm} \cdot 10^5$ [Pa]	Volume of gas injected V_{inj} [ccm]	Pressure of gas injected $p_{fill} \cdot 10^5$ [Pa]
370	5	100	1.26
370	11	100	3.15
390	8	80	2.90
410	5	60	2.32
410	11	60	5.81

Filling pressure is shown as a function of the oil volume V_{oil} [ccm] and gas pressure in the accumulator $p_{acm} \cdot 10^5$ [Pa] and satisfies the assumption that the oil is saturated with gas. The cavity volume V_{inj} [ccm] depends on the oil volume and is the difference between the total inner volume of 470 [ccm] and V_{oil} [ccm]. Next, the tooling was closed, sealed and an amount of gas was injected into the oil chamber according to Table 2, so that the entire volume of the remaining space was filled.

Lastly, the gas accumulator was filled with the appropriate amount of gas to achieve the pressure specified in Table 1. The dynamic test began after mounting such a prepared tooling on the test machine, by executing the test program.

After program completion, the tooling was dismantled from the machine and preparation for the next test run began; the same procedure was followed for each row in Table 1. Atmospheric air, compressed by a mechanical device was used to pressurize gas in both chambers, the cavity in the oil chamber as well as the gas accumulator.

5 Data Analysis

Measurement signals taken in each run of the DOE experiment were pre-processed in preparation for calculating the quantitative measure of the aeration phenomenon [8]. In the first step, the signal was divided into segments; minima of the displacement signal were assumed to be the markers of segmentation. In the second step, the cycle previous to the last one was selected for smoothing and all derivatives up to the order two of the displacement and the force signals were computed in order to obtain the following differential form

$$D(t) = \begin{vmatrix} d'(t) & F'(t) \\ d''(t) & F''(t) \end{vmatrix} = d'(t) \cdot F''(t) - d''(t) \cdot F'(t) \tag{25}$$

Zeros of D(t) indicate inflection points of the force-displacement curve. An exemplary result of the procedure is shown in Fig. 3. The smoothed curve (the so called basic shape) shown in Fig. 3a was obtained for the frame length of n = 33 points and the degree of the smoothing polynomial k = 3.

The quantitative aeration measure is derived as the ratio of the distance between the minimal displacement of the selected cycle and the value of the displacement corresponding to the convex-to-concave inflection point. In order to ensure accuracy, the authors chose to determine the inflection point manually based on tracing the plot in Fig. 3b. Here, the clock-wise direction is assumed to be the natural (positive) direction (corresponding to the positive direction of the time axis). The interior of the region enclosed by the F(d) curve is indicated by the vector pointing towards the origin.

The results of data processing are presented, along with a definition of the DOE matrix, in Table 1. Since the full factorial design was chosen for the current research, a linear model was fitted to the data. Model adequacy is confirmed by the lack of any visible structure in residuals. The factor terms are significant at a 0.95 confidence level, while the interaction term is insignificant. The reduced model takes the form

$$aeration(V_{oil}, p_{acm}) = 6.2 \cdot V_{oil} + 2.1 \cdot p_{acm} + 23.1 \tag{26}$$

The fit quality of the reduced model, indicated by the parameter $R^2 = 97.9$, is very good. The influence of V_{oil} is stronger than that of p_{acm} (Fig. 5).

The main effect plot (Fig. 5a, b) as well as the interaction plot (Fig. 5c) indicate the combination of parameters that yields the optimal solution for the problem, i.e. the solution minimizing the aeration effect is the corner point $V_{oil} = 370$ [ccm] and $p_{acm} = 5 \cdot 10^5$ [Pa], for which the fitted value of the aeration measure amounts to 14.6%. Both factors in the model are significant, with V_{oil} having a stronger effect, thus a corner point (370.5) is the optimal selection of parameters that minimizes the negative effect of oil aeration. Interaction plot confirms that the curvature in measurement data does not

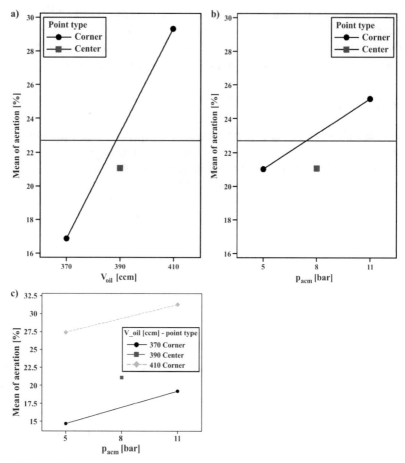

Fig. 5. Main effects plots in case of: a) volume of oil V_{oil}, b) pressure in accumulator p_{acm}, b) interactions plot.

invalidate the conclusion on the optimal factor combination. The location of the center-point value below the mean-line indicates the presence of the curvature; the curvature can, however, be disregarded as is it not significant at the 0.95 confidence level.

6   Conclusions

In this paper, a comprehensive discussion concerning the results of research into the effects of oil aeration in shock absorbers is presented. A methodology for constructing and validating a measurement system capable of quantifying such effects is provided, along with illustrative examples, which consist of the results of the performed measurements and simulation using a specially designed tooling that emulates all the working principles of shock absorbers used in practice in many branches of engineering. Moreover, the mathematical model of the phenomenon is stated, and its validity confirmed by showing the correlation between the aeration measure and the parameter χ from the simulation model (Fig. 6).

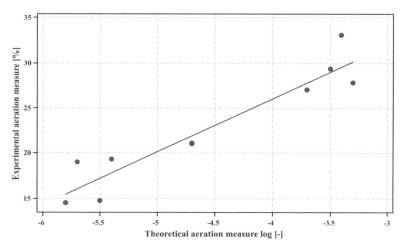

Fig. 6. Correlation plot (the Pearson correlation coefficient amounts to 94.5 whilst the linear model fit quality is $R^2 = 89.3$).

The results of the research work underlying this paper clearly show that the aeration phenomenon in shock absorbers is complex in the sense that the magnitude of the effect depends on multiple parameters, some of which are uncontrollable, or noise.

The paper shows that, firstly, the phenomenon of interest can be measured accurately by providing a data-processing algorithm and, secondly, that assumptions concerning its physics are complete and correct, by showing that the simulation results closely agree with the measurements, both qualitatively and quantitatively.

Moreover, it derives the methodology using a simple mathematical description of the solubility theory to describe complex effects caused by the fact that aerated oil is forced to flow through geometrically complicated channels, and that its flow changes the geometry itself in a way dependent on the forcing signal dynamics. The model uses this theory in a global, or averaged, way to model the loss of oil density and viscosity caused by the presence of an oil-gas-bubble emulsion as the working medium. The authors are convinced that the research presented in this paper and, more specifically, the analytical techniques employed to characterize the phenomenon, as well as the description of the measurement system and derivation of the simulation model, are all successful and enhance the overall understanding of the aeration phenomenon. The work performed provides an optimization methodology useful in solving practical problems arising in automotive engineering.

The current work provides additional benefits, as it indicates the direction in which future efforts are to be undertaken to further deepen the understanding of the time evolution of the aeration measure and to assess accuracy, repeatability and reproducibility indicators of the measurement system. In order to advance the work, means of controlling parameters which influence the dynamics of the aeration measure are to be established and the ability to reduce the variability of those is to be improved as much as possible. Performing further simulation work employing the techniques of 3-dimensional Computational Fluid Dynamics (CFD) and the Fluid-Structure Interaction (FSI) models has become possible thanks to experience gained while working on this material and will be published elsewhere.

References

1. Duym, S.W., Stiens, R., Baron, G.V., Reybrouck, K.G.: Physical modeling of the hysteretic behaviour of automotive shock absorbers. SAE International, USA (1997)
2. Dixon, J.: The Shock Absorber Handbook, 2nd edn. Wiley, Wiltshire (2008)
3. Ceccio, S.L., Brennen, C.E.: Observations of the Dynamics and Acoustics of Travelling Bubble Cavitation. J. Fluid Mech. **233**, 633–660 (1991)
4. Brennen, C.E.: Cavitation and Bubble Dynamics. Cambridge University Press, Cambridge (2013)
5. Czop, P., Slawik, D.: A high-frequency first-principle model of a shock absorber and servo-hydraulic tester. Mech. Syst. Signal Process. **25**, 1937–1955 (2011)
6. Mathews, P.G.: Design of Experiments with MINITAB. ASQ Quality Press, USA (2005)
7. Allen, P., Hameed, A., Goyder, H.: Automotive damper model for use in multi-body dynamic simulations. In: Proceedings of the Institution of Mechanical Engineers, Part D: Journal of Automobile Engineering, vol. 220, no. 9, pp. 1221–1233. Sage Journals, USA (2006)
8. Włodarczyk, T., Czop, P., Sławik, D., Wszołek, G.: Automatic shape identification by data-driven algorithms with applications to design optimizing in hydraulics. J. Transdiscipl. Syst. Sci. **16**(1), 35–43 (2012)
9. Bronshtein, I., Semendyayev, K., Musiol, G., Muehlig, H.: Handbook of Mathematics, 5th edn. Springer, Heidelberg (2007)
10. Press, W.H., Flannery, B.P., Teukolsky, S.A., Vetterling, W.T.: Numerical Recipes. The art of Scientific Computing. 3rd edn. Cambridge University Press, Cambridge (1989)
11. Wu, C.J., Hamada, M.S.: Experiments: Planning, Analysis, and Optimization, 2nd edn. Wiley, Wiltshire (2011)
12. Czop, P., Slawik, D., Sliwa, P.: Static validation of a model of a disc valve system used in shock absorbers. Int. J. Veh. Des. **53**, 317–342 (2010)

Development of Unmanned Cargo VTOL Aircraft

Roman Czyba[1]($\boxtimes$) , Zbigniew Gorol[2], Marcin Lemanowicz[3] , Mariusz Pawlak[2] , and Adam Sikora[1]

[1] Faculty of Automatic Control, Electronics and Computer Science, Department of Automatic Control and Robotics, Silesian University of Technology, Akademicka Street 2A, 44-100 Gliwice, Poland
roman.czyba@polsl.pl

[2] Faculty of Mechanical Engineering, Department of Theoretical and Applied Mechanics, Silesian University of Technology, Akademicka Street 2A, 44-100 Gliwice, Poland

[3] Faculty of Chemistry, Department of Chemical Engineering and Process Design, Silesian University of Technology, Akademicka Street 2A, 44-100 Gliwice, Poland

Abstract. The aim of the paper is to present a design cycle regarding to the construction of an unmanned cargo aircraft with own weight up to 25 kg, operating range of up to 100 km and AGL operating ceiling of up to 1 km. The task of the project team was to plan the design cycle, perform analytical calculations and CFD simulations, and finally to build an unmanned aircraft capable of lifting cargo. At the same time, a number of assumptions and technical regulations regarding constructions had to be met, which are specified in legal regulations regarding the given class of unmanned aerial vehicles. In order to achieve optimal UAV aerodynamic properties, including maximization of the payload while minimizing the own weight, aerodynamic calculations and CFD simulations were performed.

Keywords: UAV · CFD · VTOL

1 Introduction

Aviation is an area intensively regulated by national, EU and international regulations, because the use of airspace is associated with the safety of people and property, which forces standardization and certification of solutions often at the supranational level and certification of equipment and institutions involved. This applies to manned aviation and directly translates into developing unmanned aviation. Recent changes in legislation should stimulate the development of new technologies and drone services and their commercialization. There is a very intensive development of technologies around the world that will constitute in the future infrastructure in urban space, the so-called U-space [1]. Last year Polish Air Navigation Services Agency launched activities related to obtaining information about the real parameters of flight, as one of the key elements necessary not only for the development of mass market and enable autonomous flight, but also air traffic safety and the future integration of unmanned aircraft with manned aviation.

A. Mężyk et al. (Eds.): SMWM 2020, AISC 1336, pp. 31–42, 2021.
https://doi.org/10.1007/978-3-030-68455-6_3

The unmanned aerial vehicle sector is becoming the most dynamic sector in the global aviation industry. The global value of the drone market in the civilian applications segment in the period 2017–2026 is estimated at nearly USD 73.5 billion, in which Poland's share is in the order of PLN 3.3 billion [2]. The development of the unmanned aerial vehicle market in Poland, i.e. vehicles using low altitude spaces (up to 150 m), has been intense for several years [3]. Assessing the potential of professional UAV applications, it should be noted that their development depends on many advanced technologies that are still in the development phase [4–9]. Drones began to gain market thanks to the progress in many technologies, such as: plastics and composite materials, allowing to build lightweight and durable constructions, small efficient engines, integrated energy-saving digital electronics, light batteries with high efficiency, accurate sensors, efficient radio communication protocols, etc. When considering the usefulness of drones as useful tools for local communities, we take into account their practical technical capabilities in specific groups of applications. Especially during the pandemic and the existing threat of infection with the COVID-19 virus, the demand for transport services related to contactless delivery of shipments of various weights and sizes is clearly visible. This market segment is at an early stage of development, and the approach proposed in this paper is an innovative prototype solution.

The purpose of this paper is to present the entire design process and construction of an unmanned cargo aircraft. The project was performed by a group of six students supported by three tutors, as part of an individual study program implemented in the form of Project Based Learning. The goal of U-Space is to coordinate drone traffic, provide IT support and create infrastructure for UAV users and service providers. In this context, the construction of a cargo airplane, as one of the elements of the U-Space system, perfectly fits into the development trends of Industry 4.0. The unmanned cargo aircraft gives the possibility of unmanned cargo transport over considerable distances.

The paper is organized as follows. First, the introduction is provided in Sect. 1. Then, the research methodology and design assumptions are presented in Sect. 2. CAD design and mechanical construction are shown in Sect. 3. Next, the computational fluid dynamics (CFD) simulations are described together with a discussion of received results. The conclusions are briefly discussed in the last section.

2   Research Methodology and Design Assumptions

The main purpose of this paper is to present the concept of VTOL cargo airplane, from scratch to final implementation, covering the individual stages of prototyping, and ending with practical implementation. The aim of the project was to build an unmanned cargo airplane with an own weight up to 25 kg, with an operating range of up to 100 km and an AGL operating ceiling of up to 1 km. The task of the project team was to plan the design cycle, perform calculations and perform CFD simulations, and then to build an unmanned aircraft capable of lifting cargo. At the same time, a number of assumptions and technical regulations regarding constructions had to be met, which are specified in legal regulations regarding a given class of unmanned aerial vehicles. The following load requirements have been assumed: up to 1.5 kg in VTOL mode, up to 8 kg in airplane mode. In order to achieve optimal UAV aerodynamic properties, including maximization of the payload while minimizing the own weight, fluid mechanics calculations and CFD simulations were performed.

The assumed research methodology is based on the rapid prototyping method and includes the following phases:

– development of detailed project assumptions,
– designing the airframe model in the Inventor environment,
– performing CFD simulations in the Ansys environment,
– airframe optimization based on CFD analysis,
– design of the structural part of the airframe,
– molds preparation,
– airframe construction,
– conducting experimental studies and flight tests.

In this paper, a novel design configuration in a form of hybrid of conventional aircraft and quadrotor, with two of four rotors mounted in nacelles with ability of rotation has been proposed (Fig. 1).

Fig. 1. VTOL aircraft configuration.

It will have a unique feature among similar solutions - a thrust vectoring of engines that ensures their usage in both vertical and horizontal flight. During takeoff, landing and hover, this solution makes that the aircraft is more robust to cross winds. The innovative solution also applies to the fuselage structure with built-in cargo compartment and transport container.

3   CAD Design and Mechanical Construction

3.1   Fuselage

One of the main assumptions of the cargo plane construction project was to optimize the fuselage VTOL aircraft so that it was able to carry a load of $200 \times 150 \times 100$ mm.

In addition, it was decided that the fuselage would be made in full composite technology so as to maximize its aerodynamic and strength properties. To fit the required load, the fuselage at the center of mass has been widened and flattened at the bottom to allow the load container to be accommodated (Fig. 2). After obtaining a satisfactory shape, the fuselage internal structure project began. The first step was to consider the inspection holes that are necessary for equipment assembly and periodic maintenance of the aircraft. Four main cut-outs were designed (Fig. 2): battery compartment opening (fuselage top), electronics compartment opening (fuselage back), cargo compartment (fuselage bottom) and Navigation systems compartment (nose).

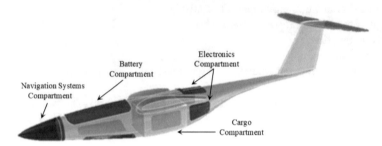

Fig. 2. Fuselage with cargo compartment and inspection cut-outs.

The entire fuselage structure is based on a sandwich type structure - it is constructed by placing foam core between the layers of carbon fabric to ensure rigidity. To further strengthen the fuselage, the structure also includes unidirectional carbon fabric straps (Fig. 3).

Fig. 3. Arrangement of carbon belts inside the fuselage.

After developing the structure, it was possible to start designing negative molds (Fig. 4) that were used to produce the correct fuselage shape using the vacuum-forming method. One half of the form consisted of 3 elements - tail, inner and outer. The milling process consisted of 2 milling strategies - rough milling (focused on removing as much material as possible in the shortest possible time) and fine milling of the final shape. The entire molds preparation process was time consuming and involved milling as well as multiple impregnation, polishing and waxing. The final stage was painting the forms in color, which was then transferred to the fuselage in the manufacturing process.

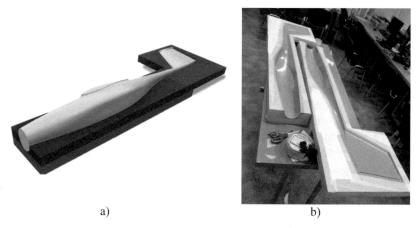

a)   b)

Fig. 4. Fuselage mold: a) design, b) final realization.

Laminating itself is the final stage in the production of the aircraft. The fuselage halves were laminated separately, and then glued together with strips of carbon fabric. Lamination started of laying fabrics and foams inside the mold and simultaneously impregnated it with resin. Then an polyamide fabric was applied as a removable layer to reduce the weight of the product. At the end of the process, the mold was closed inside a vacuum bag to ensure even pressure against the mold surface.

After the resin had hardened, the fuselage was removed from the mold (Fig. 5).

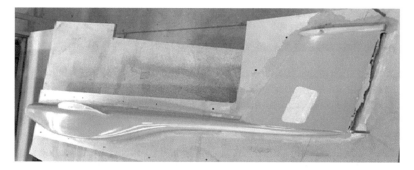

Fig. 5. Fuselage during demoulding.

3.2 Wing

The most important element of any airframe are wings. They must provide not only the ability to transfer load over their entire surface, during horizontal flight, but provide sufficient rigidity and strength during hovering, where the force affects them almost point-wise. At the production stage, the same technology was adopted as the fuselage - sandwich type structure. The first difference was the use of the ultra-light Textreme fabric, which, when correctly laid and laminated, is characterized by exceptional torsion resistance (Fig. 6). Another difference in laminating relative to the fuselage was the use of unidirectional carbon roving whose quantity and distribution was calculated for this particular wing, in order to obtain a graduated girder, adapted to the distribution of forces on the wing during flight. This treatment significantly reduced the weight of the wing. The last difference in performance was the use of special polyamide fiber in places where the control surface hinge was to be created. During the processing of the wings after removal from the mold, delicate undercutting of the surface in the places provided for the hinges allowed to obtain an excellent aerodynamic hinge. In the final phase, elongated pipes were glued into the intended places, which constituted the fixing points of the propulsion system.

Fig. 6. Wing lamination process.

3.3 Engine Nacelles and Fairings

At the initial design stage, the dimensions of the engine nacelles and fairings necessary to fix them inside were determined, and what effect have to be achieved by using fairings. The fairing surface is covering all the elements that could have affected the aerodynamics of the aircraft. To achieve the lowest possible weight, the fairing was divided into two parts (Fig. 7).

The first part is covering of the front engine and its mounting, and the second one is covering of the rear engine and its mounting. The shape should be as aerodynamic as possible, similar to the shape of the wing profile. The frontal surface has been reduced and the tips profiled. Finally, the cross section of the front and rear fairings was rounded. In addition, a special hole has been designed on the front face of the fairing to allow free movement of the rotation mechanism. The rear fairing, on the other hand, has been minimized to cover only the rear engine mount (Fig. 7).

Below are CAD drawings of the initial concept and the final concept that meets the requirements.

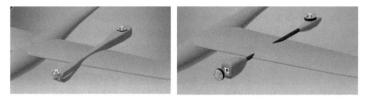

Fig. 7. First (left) and final (optimal) (right) concept of engine nacelles with fairings (front and rear).

The manufacturing process consisted of the following stages:

a) Preparation of negative fairing models (Fig. 8a) and preparation of G-code for milling machine, using Inventor software,

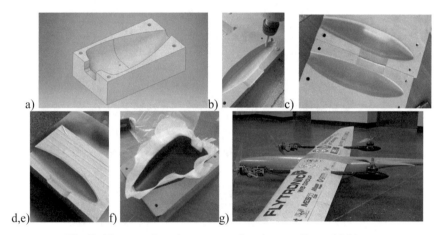

Fig. 8. The manufacturing process of engine nacelles and fairings.

b) Mold milling consisting of two operations: rough milling and finishing (Fig. 8b),
c) Grinding the forms with sandpaper with less and less gradation and polishing (Fig.8c),
d) Waxing of molds - application of 8 layers of wax to prevent sticking of manufactured elements to the mold (Fig. 8d),
e) Painting molds using matte aerosol paint (Fig. 8e),
f) Lamination using 3 layers of fiberglass - 40 g/m^2 (Fig. 8f).

3.4 Cargo Compartment

One of the tasks that the aircraft was to perform was the ability to transport cargo. It was assumed that the aircraft should be able to lift at least 1–1.5 kg of cargo, while its dimensions can reach up to $100 \times 150 \times 200$ mm. The location and orientation of the loading chamber with the assumed dimensions and mass depended on the aerodynamic properties of the aircraft and the location of the assumed mass center. As cargo weighing 1–1.5 kg was to be a significant part of the maximum take-off weight, and it should also be assumed that this value is variable (the aircraft can carry a lighter load or not at all), the chamber must be located near the center of gravity and as close as possible to the aircraft's supporting surfaces. Therefore, it was decided to place the loading chamber in the hull, under the wings. The hull has been specially designed and widened in its lower part to be able to contain 150 mm wide cargo. Because of the limited space and the need to make sure that the chamber does not cross the plane completely, separating the front from the rear, the smallest dimension - 100 mm as the height. The longest dimension - 200 mm - was oriented along the hull. The overall dimensions of the whole chamber should be assumed larger than the load that is to be contained in it, which was one of the design problems. Planned location of the cargo on the aircraft is show in Fig. 9.

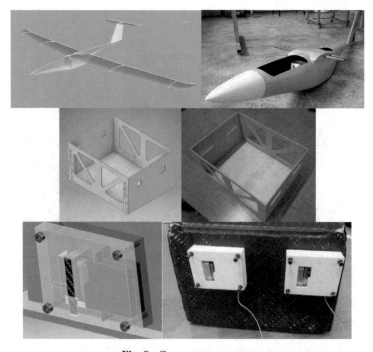

Fig. 9. Cargo compartment.

As part of the design work, a free space was created in the model of the aircraft body and a box with a trigger system was designed. Several concepts were created, the final version of the box with the electro-mechanical charge release system is presented on Fig. 9. The loading chamber was made in composite technology, similar to the hull. The loading case is made of balsa plywood. To minimize the space needed to lift the box, an improved ratchet design has been developed. Its structure is now included in the modules mounted to the chamber as a whole, and the springs have been replaced with flexible plates. The CAD design of this module is also shown in Fig. 9.

4   CFD Analysis

In order to determine the aerodynamic properties of the designed platform a series of Computational Fluid Dynamics (CFD) simulations had to be done. For the calculations ANSYS Workbench 19.2 environment was used with Fluent software as the CFD solver.

The CAD geometry was imported to the SpaceClaim software where it was appropriately prepared for the mesh generation. The UAV enclosure was created on the basis of paraboloid. The origin of the coordinate system was placed at the center of gravity of the VTOL model. The domain extended in the range of -15 m to 30 m in X direction and -15 m to 15 m in Y and Z directions. However in order to reduce the computational cost of performed simulations the symmetry plane was introduced by slicing the paraboloid in half in the XZ plane, thus reducing the number of cells by half. Next the grid independence study was carried out for number of cells varying from 5 mln up to 15 mln. As the optimization criterion the value of the lift force was used. The series of unstructural hybrid meshes were generated. The optimal mesh was composed of 9 mln elements and it was characterized by the following quality parameters: *minimal orthogonal quality 0.013, maximal skewness 0.985, maximal aspect ratio 644.* As one should expected the cells of poorest quality were placed in the inflation layer. Yet it was essential for the proper development of the velocity gradient at the solid walls of the model and thus for the valid calculations of forces acting on the plane.

As the CFD solver ANSYS Fluent 19.2 was used. Since relatively low velocities were supposed to be simulated the pressure based solved was chosen. Moreover constant physicochemical parameters of air were assumed. In order to calculate the turbulent viscosity the *k-omega SST* turbulence model developed by Menter [10] was used. It effectively blends the *k-omega* model in the vicinity of the solid walls with *k-epsilon* model in the far field. In the result, a robust and accurate prediction of turbulence is provided for the whole domain. The transport equations for the *k-omega SST* model take the form [11]:

$$\frac{\partial}{\partial t}(\rho k) + \frac{\partial}{\partial x_i}(\rho k u_i) = G_k^* - Y_k^* + S_k^* + \frac{\partial}{\partial x_j}\left(\Gamma_k \frac{\partial k}{\partial x_j}\right) \tag{1}$$

$$\frac{\partial}{\partial t}(\rho \omega) + \frac{\partial}{\partial x_j}(\rho \omega u_j) = G_\omega - Y_\omega + D_\omega + S_\omega + \frac{\partial}{\partial x_j}\left(\Gamma_\omega \frac{\partial \omega}{\partial x_j}\right) \tag{2}$$

The turbulent viscosity is calculated using [8]:

$$\mu_t = \frac{\rho k}{\omega} \frac{1}{\max\left[\frac{1}{\alpha^*}, \frac{SF_2}{a_1 \omega}\right]} \tag{3}$$

where F_2 is the blending function.

As for the boundary conditions, the side surface of the paraboloid was set as *velocity inlet*, the XZ surface was set as symmetry plane and the base of the paraboloid was set as *pressure outlet*. The pressure-velocity coupling was realized via coupled scheme, the gradients were calculated using the Green-Gauss Node Based Method whereas the spatial discretization of pressure was done using second order scheme and the momentum, turbulent kinetic energy and specific dissipation rate using second order upwind schemes.

Finally pseudo-transient calculations were enabled in order to facilitate the convergence of simulations. The convergence criteria were set to 1E−03 for each parameter.

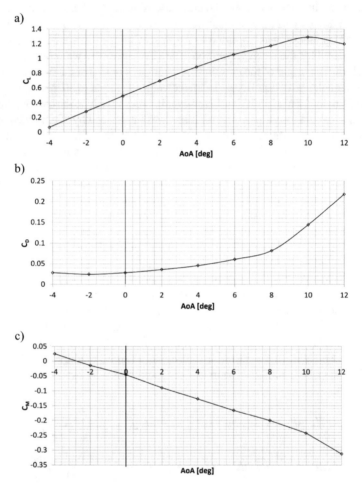

Fig. 10. Aerodynamic loads: a) lift coefficient vs angle of attack b) drag coefficient vs angle of attack c) Pitching moment coefficient vs. angle of attack.

The simulations were made for the angle of attack range from −4° up to 12° and the velocity equal to 16 m/s. In Fig. 10 the lift coefficient, drag coefficient and pitching moment coefficient are presented respectively. They were calculated for the reference area equal to 0.6726 m² and chord length equal to 0.285 m. The simulations proved that the designed UAV has satisfactory aerodynamic characteristics. For the given velocity (which was considered as the minimal allowed flight velocity) the lift generated by the construction surpass the critical value. The maximal lift was achieved for 10° and it was equal to 136 N. The highest lift to drag ratio was achieved for the 4° angle of attack and it was equal to 19.5. The stall phenomenon occurred for angle of attack beyond 10°.

Simultaneously, the drag generated by the plane is relatively low. These features together with negatively sloping pitching moment coefficient will provide stable behavior of the VTOL during horizontal flight.

Moreover, the reserve of lift force will allow one for the installation of additional equipment thus expending the capabilities of the plane without negative impact in its maneuverability od flight range.

Figure 11 present the contour graphs of the velocity around the VTOL wing for 0° and 10° angle of attack respectively. One may clearly notice the phenomenon of air stream detachment from the wing surface resulting in stall phenomenon. Although the range of the angle of attack for the horizontal flight is sufficient this process may be an issue during the transitional stage of flight. Therefore a special care has to be taken for the development of the VTOL control algorithm.

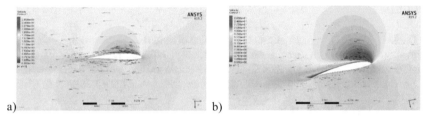

Fig. 11. Velocity contour profile for a) AoA $= 0^\circ$ and Re $= 320\,000$ b) AoA $= 10^\circ$ and Re $= 320\,000$.

5 Conclusions

The purpose of this paper was to present the entire design process and construction of an unmanned cargo aircraft. The final effect of the project is a VTOL cargo aircraft with a hybrid design that combines a classic airframe and a rotor platform. The specificity of vertical take-off and landing makes it possible to maneuver in a small space and use the aircraft in the urban space. The innovative solution also applies to the fuselage structure with built-in cargo compartment and transport container. As a part of Project Based Learning, the student team gained extensive knowledge on the practical aspects of designing and producing unmanned airframes in laminate technology.

The solution proposed in this article may offer significant market benefits. For transport and logistics, this means the development of strictly dedicated supplies for a specific recipient, i.e. a reduction in the movement of heavier vehicles, which in turn translates into benefits for environmental protection. It also gives the opportunity to organize immediate shipments for consumers, and on a larger scale to reduce the cost of handling retail delivery. However, one should take into account, among other things, the necessity to introduce regulations in the area of flight corridors, preparation of landing pads, adaptation of the city's spatial planning system to the third dimension, ensuring access to buildings, defining prohibited areas.

There is a very intensive development of technologies around the world that will constitute U-space infrastructure in the future. Such solutions exist and are being tested, but it is difficult to say that the market for U-space infrastructure solutions has already emerged. On the contrary, it is still immature and not standardized, and individual countries are just beginning to test various solutions. However, looking at the dynamics of such area, it can be safely said that the first significant implementations will begin to appear soon.

Acknowledgment. The research reported in this paper was co-financed by the European Union from the European Social Fund in the framework of the project "Silesian University of Technology as a Center of Modern Education based on research and innovation" POWR.03.05.00-00-Z098/17.

References

1. Doole, M., Ellerbroek, J., Hoekstra, J.: Estimation of traffic density from drone-based delivery in very low level urban space. J. Air Transp. Manage. (2020). Article number 101862
2. Darowska, M., Kutwa, K.: Biała Księga Rynku Bezzałogowych Statków Powietrznych. Polski Instytut Ekonomiczny, Ministerstwo Infrastruktury, Warszawa (2019). ISBN 978-83-61284-74-1
3. Konert, A., Kotlinski, M.: Polish regulations on unmanned aerial vehicles. In: 7th International Conference on Air Transport (INAIR), Hainburg an der Donau, Austria, 20–21 November 2018
4. Perz, R., Wronowski, K.: UAV application for precision agriculture. Aircr. Eng. Aerosp. Technol. **92**(20), 257–263 (2019)
5. Austin, R.: Unmanned Aircraft Systems –UAVS Design, Development and Deployment. Wiley, New York (2010)
6. Valavanis, K.P.: Advances in Unmanned Aerial Vehicles. Springer, Dordrecht (2007)
7. Castillo, P., Lozano, R., Dzul, A.E.: Modelling and Control of Mini-flying Machines. Springer, London (2005)
8. Nonami, K., Kendoul, F., Suzuki, S., et al.: Autonomous Flying Robots, 1st edn. Spirnger, Tokyo (2010)
9. Saeed, A.S., Younes, A.B., Islam, S., Dias, J., Seneviratne, L., Cai, G.: A review on the platform design, dynamic modeling and control of hybrid UAVs. In: International Conference on Unmanned Aircraft Systems (ICUAS), Denver, Colorado, USA, pp. 806–815 (2015)
10. Menter, F.R.: Two-equation Eddy-viscosity turbulence models for engineering applications. AIAA J. **32**(8), 1598–1605 (1994)
11. ANSYS Fluent Theoretical Guide, ver. 19.2

Model Tests of the Effect of the Column Centre of Gravity Position on the Value of the Passive Safety Coefficient ASI During Impact

Wojciech Danek[(✉)] [iD], Mariusz Pawlak [iD], and Damian Gąsiorek [iD]

Institute of Theoretical and Applied Mechanics, Silesian University of Technology, Gliwice, Poland
wojciech.danek@polsl.pl

Abstract. This paper shows the effect of mass on passive safety coefficient (Acceleration Severity Index - ASI) during vehicle collision with a lighting column. The subject related to reduction of passive safety coefficient was chosen because every year in Europe number of fatal accidents is around 25 000. Part of this fatal accident are with poles and signs constitute about 1.5–2% of all fatal accidents. Proposed approach allowing reduction of passive safety coefficient concerned the influence of the position of the lighting column's center of gravity on the previously mentioned coefficient. For this purpose, a test stand was created allowing a change in the position of the center of gravity of the lighting column.

Keywords: Car crash · Fast camera · Lighting column · Passive safety

1 Introduction

With the development of technology and production techniques, new concepts and ideas regarding the development of space around roads are being created. Road installations in the field of potential contact with the vehicle must meet increasingly higher safety and aesthetic requirements. These constantly growing requirements are connected with adopted by the governments of various countries safety programs [1–3]. They are aimed at reducing the number of fatal road accidents, which in recent years have been around 25,000 [4]. In the case of lighting columns, the division into the foundation and the part directly mounted in the ground is visible, where the first of them have a problem with too high ASI indicator, the second one with THIV. In spite of the increasing use of composite materials for lighting columns, a large proportion of aluminum constructions are characterized by relatively high resistance to weather conditions. The influence of the car's center of mass on the value of the safety coefficient was analyzed by the authors in publications [5–7], so in order not to rethread the material, the influence of the center of mass of the lighting column with the equipment on top and what is in the ground (foundation) will be analyzed.

There are many scientific publications related to car crash with road infrastructure [10–19]. Car crash with roadside barriers were described by authors in publications

© The Author(s), under exclusive license to Springer Nature Switzerland AG 2021
A. Mężyk et al. (Eds.): SMWM 2020, AISC 1336, pp. 43–52, 2021.
https://doi.org/10.1007/978-3-030-68455-6_4

[10–12]. Authors in this publications show results of numerical simulation and experimental research with different types of roadside barriers. Other elements of road infrastructure like crash cushion or reinforced concrete bridge columns are also subject of the publications [13, 14]. In article [16] the authors presented the influence of the lighting column wall thickness on the acceleration value and the level of deformation of the vehicle during the collision obtained on the basis of a virtual crash test. Article [17] show influence of the composite lighting column height on the values of ASI and THIV coefficient. Based on the literature review about car crash with lighting column can be noticed that there are not research about influence of mass parameters of the value of passive safety coefficient.

2 Test Stand

In order to verify the assumed hypothesis about the influence of the center of the mass position of the lighting column on the passive safety coefficients, tests were carried out using a research stand designed to simulate the collision of the vehicle with the lighting column. The research stand was created as part of the MA thesis written at the Institute of Theoretical and Applied Mechanics. The design of the test device based on the Charpy hammer, which allows testing of materials' resistance to impulse excitation. Its scheme is presented in Fig. 1.

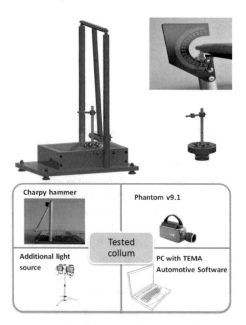

Fig. 1. Visualization of the research stand with equipment [5].

In order to determine kinematic parameters and ASI coefficient occurring during the collision, fast camera Phantom v9.1 was used. This camera allowed to record movies with the frequency equal to 150000 Hz. Motion Analysis was conducted using TEMA Automotive software. In the case of tests carried out on the testing device, the recording frequency was 3500 Hz.

The list of measuring points is presented below (Fig. 2)

- Point 1 – a point on the surface of the pendulum,
- Point 2, 3 – points enabling the determination of the scale and conversion of values from pixels to SI units,
- Point 4 – a point on the surface of the column

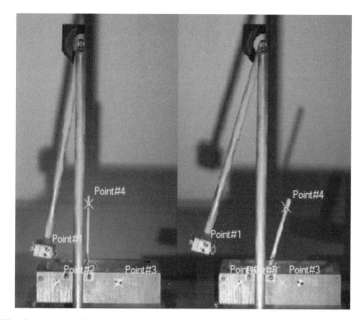

Fig. 2. Scheme of the measurement station with marked measuring points [5].

Such a way of distribution of the measurement points made it possible to determine the value of ASI passive safety coefficient, impact energy value, and the behavior of the column as well as pendulum during the collision. Aluminum was chosen as the material used for the column in model tests.

The impact energy was determined based on the knowledge of the pendulum's mass and the impact velocity determined on the basis of the analysis of the recorded film and its average values were equal to:

- E1 = 9.15 J
- E2 = 11 J
- E3 = 13 J

Three ways of introducing additional mass to the element allowing for the change of position of the gravity center were analyzed. The first method involves adding the mass to the upper part of the lighting column. This procedure resulted in the gravity center of the examined element being raised but still located on the axis of the element. The second method involves adding the mass in a fragment of the column located underground. This caused the gravity center to move lower but still in the axis of the element. The third method involves adding mass on the bracket. This resulted in both moving the gravity center up and simultaneously away from the axis towards the bracket. Illustrations showing the location of the center of mass for each case are presented in Fig. 3.

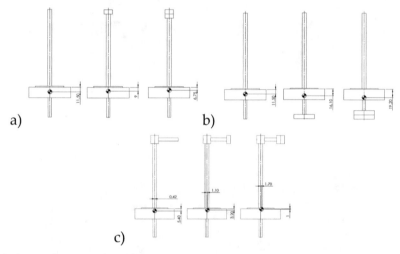

Fig. 3. Research stand while adding a) mass in upper position b) mass in lower position c) mass on the bracket.

3   Results

For each of the methods of additional construction load, tests were carried out for three different impact energy values, designed to simulate the impact of vehicles with different weights. The values of the ASI coefficient during the collision in the case of moving the gravity center position higher were presented in Fig. 4 and 5

Based on the data presented in Fig. 4, it can be observed that shifting the gravity center position up reduces the value of the ASI passive safety coefficient. Raising the position of the center of mass by 2.5 mm causes a decrease of the value of ASI coefficient about 9%, whereas if the gravity center was raised by 4.75 mm, the value of ASI coefficient decreased about 13% in relation to the initial value. Analyzing the data presented in Fig. 5, for a solution where the gravity center position was raised by 4.75 mm, it was observed that the increase in impact energy causes the increase of the ASI coefficient value. The values of the ASI coefficient during the collision in the case of moving the gravity center position lower were presented in Fig. 6 and 7

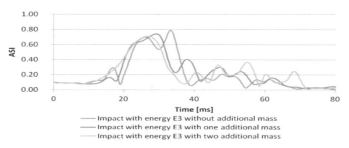

Fig. 4. The effect of moving the gravity center higher in the column axis on the value of the ASI coefficient.

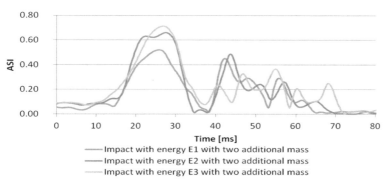

Fig. 5. The effect of impact energy in the case of moving the gravity center higher in the column axis to the value of the ASI coefficient.

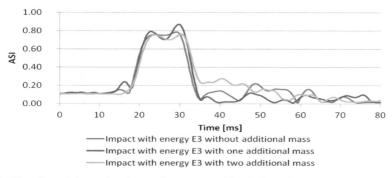

Fig. 6. The effect of decreasing the gravity center position in the column axis on the value of the ASI coefficient.

Based on the data presented in Fig. 6, it can be observed that lowering the gravity center by 4.6 mm increases the value of passive safety coefficient by about 14%, while lowering gravity center by 7.7 mm the value of this coefficient decreases by about 1%. Analyzing the data presented in the Fig. 7, for the solution where the gravity center was lowered by 7.7 mm, it was observed that an increase in impact energy above a certain

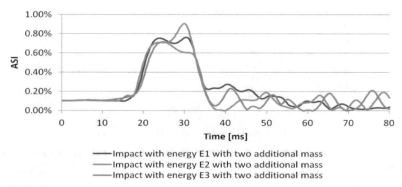

Fig. 7. The effect of impact energy for the case of decrease the gravity center in the column axis to the value of the ASI coefficient.

value causes a decrease in the ASI value. The values of the ASI coefficient during the collision in the case of moving the gravity center position beyond axis of symmetry were presented in Fig. 8 and 9

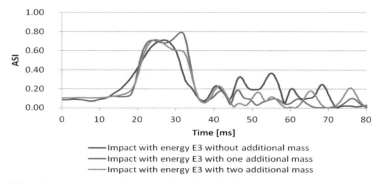

Fig. 8. Effect of moving the gravity center beyond axis of symmetry of the column on the value of the ASI coefficient.

Based on the data presented in Fig. 8, it can be observed that moving the position of the gravity center from the axis of symmetry of the column does not affect the decrease of the ASI coefficient. However, it is possible to observe the influence of the impact energy value on the value of this coefficient, where with the E3 energy, the decrease of the ASI value by about 6% was observed.

The columns behavior during the impact were shown in Figures from 10 to 11

On the basis of the graph presented in Fig. 10, it can be observed that the raising the center of mass, achieved by adding the mass to the upper part of the column, makes it possible to reduce the maximal deflection of the lighting column. Regarding the influence of the impact energy on the columns deflection, it can be seen that as the impact energy increases, the maximal column deflection increases. In addition, for the highest impact energy for the case without additional mass, after the expansion ceases, the column only

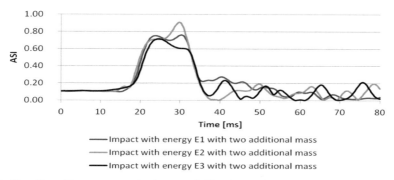

Fig. 9. The effect of impact energy for the case of offset the gravity center beyond axis of symmetry of the column on the value of the ASI coefficient.

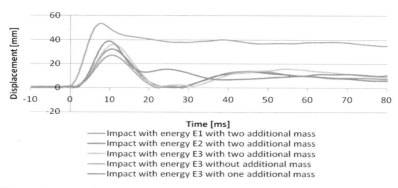

Fig. 10. Displacement of the top of the column (point 4) during the impact for the case of raising the center of mass.

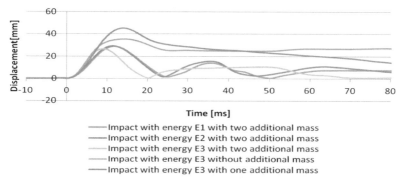

Fig. 11. Displacement of the top of the column (point 4) during the impact for the case with lowered center of mass.

slightly returned to its initial position, where for the remaining cases the position of the column after impact was near the starting point.

Analyzing the data presented in Fig. 11, we can observe the effect of lowering the center of mass on the displacement of the top of the column. The highest value of the lighting column deflection was achieved in the impact with energy E3 with one additional load, while the smallest inclination was achieved in the impact with energy E3 with two additional loads. Based on this data, it can be concluded that the decrease of the columns deflection during the impact occurs only after adding additional mass causing the gravity center to decrease. For cases of impact with energy E3 with one and two additional loads, after the impact column did not return to the initial position as it was in other cases (Fig. 12).

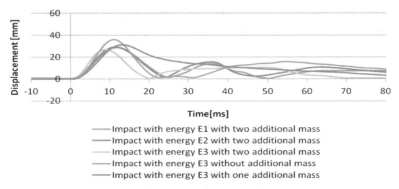

Fig. 12. Displacement of the top of the column (point 4) during the impact in the case of moving the gravity center beyond axis of symmetry of the column.

If the position of the gravity center is moved from axis of symmetry of the column, it can be seen that this parameter does not affect the displacement of the top of the column during the impact like in the case of raising or lowering the position of the center of mass. The highest displacement of the top of the column was observed in impact with E3 energy without additional loads, while the lowest in impact with E3 energy with two additional loads.

4   Conclusion

Based on the results obtained from experimental research on test stand it can be concluded that increase position of center of the gravity allowing reduction of passive safety coefficient ASI more than 10%. In case of decreasing position of the center of gravity it was observe that an decreasing position of center of gravity above a certain value causes a decrease in the ASI value. For situation where center of gravity was translate from the axis of symmetry of the column it was not observe decreasing of the ASI coefficient. In future works authors plan to make some modification allowing for a better representation of the impact conditions. The phenomenon obtained from a small stand due to the different size may be difficult to reflect on a real object.

References

1. 5th EU Road Safety Action Programme 2020–2030. https://etsc.eu/wp-content/uploads/5th_rsap_2020-2030_etsc_position.pdf
2. Road Safety Programme 2011-Federal Ministry of Transport, Building and Urban development, Germany. https://ec.europa.eu/transport/road_safety/sites/roadsafety/files/pdf/20151210_2_germany_road-safety-programme-2011.pdf
3. Narodowy Program Bezpieczeństwa Ruchu Drogowego 2013–2020, Krajowa Rada Bezpieczeństwa ruchu Drogowego, Warszawa (2013). www.krbrd.gov.pl/files/file/NP-BRD-2020_przyjety_przez_KRBRD.pdf
4. Annual Accident Report (2018). https://ec.europa.eu/transport/road_safety/sites/roadsafety/files/pdf/statistics/dacota/asr2018.pdf
5. Danek, W., Pawlak, M.: Charpy impact testing machine in modeling of vehicle frontal crash with street lights. In: Dynamical Systems in Applications. DSTA 2017 Springer Proceedings in Mathematics & Statistics; Awrejcewicz J.; Publisher: Springer, Cham, vol. 249, pp. 73–84 (2018)
6. Pawlak, M.: Comparison of acceleration severity index of vehicle impacting with permanent road equipment support structures. In: 13th Conference on Dynamical Systems Theory and Application DSTA, Łódź, Poland 7–10.12.2015
7. Pawlak, M.: The acceleration severity index in the impact of a vehicle against permanent road equipment support structures. Mech. Res. Commun. **77**, 21–28 (2016)
8. Mężyk, A., Klein, W., Pawlak, M., Kania, J.: The identification of the vibration control system parameters designed for continuous miner machines. Int. J. Non-Linear Mech. **91**, 181–188 (2017)
9. Mężyk, A., Klein, W., Fice, M., Pawlak, M., Basiura, K.: Mechatronic model of continuous miner cutting drum driveline. Mechatronics **37**, 12–20 (2016)
10. Gutowski, M., Palta, E., Fang, H.: Crash analysis and evaluation of vehicular impacts on W-beam guardrails placed on sloped medians using finite element simulations. Adv. Eng. Softw. **112**, 88–100 (2017)
11. Li, N., Fang, H., Zhang, C., Gutowski, M., Palta, E., Wang, Q.: A numerical study of occupant responses and injuries in vehicular crashes into roadside barriers based on finite element simulations. Adv. Eng. Softw. **90**, 22–40 (2015)
12. Atahan, A.O., Arslan, T., Sevim, U.K.: Traffic safety at median ditches: steel vs. concrete barrier performance comparison using computer Simulation. Safety **4**(4), 50 (2018)
13. Büyük, M., Atahan, A.O., Kurucuoğlu, K.: Impact performance evaluation of a crash cushion design using finite element simulation and full-scale crash testing. Safety **4**, 48 (2018)
14. Abdelkarim, O.I., ElGawady, M.A.: Dynamic and static behavior of hollow-core FRP-concrete-steel and reinforced concrete bridge columns under vehicle collision. Polymers **8**, 432 (2016)
15. Munyazikwiye, B.B., Vysochinskiy, D., Khadyko, M.G., Robbersmyr, K.: Prediction of vehicle crashworthiness parameters using piecewise lumped parameters and finite element models. Designs, **2**(4), 43 (2018)
16. Abdel-Nasser, Y.: Frontal crash simulation of vehicles against lighting. Alex Eng. J. **52**(3), 295–299 (2013)
17. Borokov, A., Klyavin, O., Michailov, A., Kemppinen, M., Kajatsalo, M.: Finite element modeling and analysis of crash safe composite lighting columns, contact- impct problem. In: 9th International LS-DYNA Users Conference, Dearborn, Michigan, USA 4–6.06.2006 (2006)
18. Pawlus, W., Robbersmyr, K.G., Karimi, H.R.: Mathematical modeling and parameters estimation of a car crash using data-based regressive model approach. Appl. Math. Model. **35**, 5091–5107 (2011)

19. Wach, W.: Crash against mast according to EN 12767 standard – uncertainty of passive safety indexes calculated in programs for simulation of vehicle accidents, In: 21th Annual Congress of the European Association for Accident Research and Analysis, Brasov, Romania, 27–29.09.2012 (2012)
20. Duda, S., Gembalczyk, G., Świtoński, E.: Computational optimization and implementation of control system for mechatronic treadmill with body weight support system. J. Theor. App. Mech.-Pol. **56**(4), 1179–1191 (2018)
21. Zhao, S., Zhu, Z., Luo, J.: Multitask-based trajectory planning for redundant space robotics using improved genetic algorithm. Appl. Sci. **9**, 2226 (2019)
22. Chui, K.T., Lytras, M.D.: A Novel MOGA-SVM multinomial classification for organ inflammation detection. Appl. Sci. **9**, 2284 (2019)
23. Mittelstedt, M., Hansen, C., Mertiny, P.: Design and multi-objective optimization of fiber-reinforced polymer composite flywheel rotors. Appl. Sci. **8**(8), 1256 (2018)
24. European Committee for Standardization, EN 12767 Passive safety of support structures for road equipment—Requirements and test methods. Draft proposal of revised EN 12767. 2005–06-02. 30p.28
25. European Committee for Standardization, SFS-EN 40-3-1 Lighting Columns, Part 3-1: Design and Verification–Specification for Characteristic Loads, Edita Oyj, Helsinki, (2000)
26. European Committee for Standardization, SS-EN 40-3-2 Design and verification - Verification by testing, Svensk Standard (2013)
27. European Committee for Standardization, SFS-EN-40-3-3 Lighting Columns–Part 3–3: Design and Verification–Verification by Calculation, Finnish Standards Association, Helsinki (2004)
28. European Committee for Standardization, SFS-EN 1317-1 Road Restraint Systems, Part 1: Terminology and General Criteria for Test Methods, Edita Oyj, Helsinki, (1998)
29. European Committee for Standardization, SFS-EN 1317-2 Road Restraint Systems, Part 2: Performance Classes, Impact Test Acceptance Criteria and Test Methods for Safety Barriers, Edita Oyj, Helsinki (1998)
30. European Committee for Standardization, SFS-EN 1317–3 Road Restraint Systems, Part 3: Performance Classes, Impact Test Acceptance Criteria and Test Methods for Crash Cushions, Edita Oyj, Helsinki (2001)

Implications of the Lifting Method on Stress Distribution in the Conveyor Belt

Sławomir Duda⊕, Sławomir Kciuk⊕, Tomasz Machoczek⊕,
and Sebastian Sławski$^{(\boxtimes)}$⊕

Silesian University of Technology, 44-100 Gliwice, Poland
sebastian.slawski@polsl.pl

Abstract. Nowadays, conveyor belts with synthetic fabric made of multilayered composite materials are used in many industries. The dimensions of these devices are usually large, which makes maintenance work more difficult. Install or dismantle of conveyor belts in these devices is carried out sporadically. Because of that belt assembly devices are often overlooked during the construction process of the conveyor. This means that conveyor belts are often lifted through various types of chains or steel profiles. Due to the length of the belts, their weight is often very high. This may result in conveyor belts being damaged during install or dismantle. The paper focused on analyzing the stress distribution in composite conveyor belt depending on the way how it is lifted. Numerical analysis was performed in commercially available software based on finite element method. Two cases of lifting a belt with a weight of 2.5 tons have been compared. Based on the obtained results, recommendations were made on how to install or dismantle conveyor belts in large-sized conveyor.

Keywords: Conveyor belt · Multilayered composite · Finite element method · Vertical conveyor · Lifting techniques

1 Introduction

Conveyor belts made of composite materials are commonly used in various industries such as mining or construction. To guarantee the proper strength of the conveyor belts, they are reinforced with continuous fibers which are made from various materials. Conveyor belts made from composite materials could be divided into the three layers. Top and bottom layers are made of rubber. This layers are mainly responsible for protection of the middle layer before mechanical and chemical damages. Middle layer of the belt consist of the reinforcing layers. Reinforcing layers are made from the fabrics which are made from continuous fibers. Those layers are responsible for load transfer. Depending on the application of the belt, different number of reinforcement layers are used. The strength of the conveyor belts depends largely on the material of the used reinforcing fibers and their weave in the reinforcing layers [1]. Proper adhesion between rubber and textile layers is also important. Influence of the temperature and cure time on the adhesive strength is presented in the paper [2]. Some dynamic characteristics

© The Author(s), under exclusive license to Springer Nature Switzerland AG 2021
A. Mężyk et al. (Eds.): SMWM 2020, AISC 1336, pp. 53–60, 2021.
https://doi.org/10.1007/978-3-030-68455-6_5

of conveyor belts made from various materials are presented in paper [3]. Composite belts are often exposed to various mechanical damage during operation. Therefore, their outer surfaces are covered with rubber, which protects reinforcing fibers from various mechanical damages and environmental impact. While operating conveyor belts, it is important to avoid damaging their structure. Damage formed in the conveyor belt as a result of sharp end hammer impact is presented in the paper [4]. Change of physical and mechanical properties of dynamically damaged conveyor belt is presented in the paper [5]. Due to the application of the conveyor, the length of the belts may be large. Large length means large weight of the belt in particular, in cases where transport buckets are attached to the belts. The weight of the conveyor belt can then reach up to several tons. Due to construction of the conveyor there is not too much space to use devices which will allow safe install or dismantle of the belt, especially in vertical conveyor. Belts are often picked up using elements such as chains or steel profiles. Examples of different ways of belts lifting are shown in Fig. 1.

a)   b)

Fig. 1. Examples of conveyor belt lifting: a) lifting with use of transport belts [6]; b) lifting with use of flat bar [7].

Stress distribution in conveyor belt during install or dismantle is different, depends from the element which is used to lift it up. Improper lifting of composite belts can damage their structure.

The research presented in the paper compares the stress distribution in the conveyor belt depending on the element which is used to lift it up during install or dismantle. In the paper created models are shown. The results of numerical studies conducted with use of the finite element method are presented. Obtained results are discussed and compared. Finally, based on the obtained results the authors have formulated conclusions.

2   Numerical Studies

Numerical studies concerning on the effect of the lifting method of the conveyor belt on its stress distribution were carried out in commercially available LS-Dyna software. This software is based on the finite element method. Created models consisted of three elements: 9 mm thick conveyor belt with 350 mm width, an element used for lifting (chain or pipe) and a drum with a diameter of 670 mm, on which the belt is tensioned during the operation. In both cases, the drum was treated as non-deformable one. It is assumed that the coefficients of static and dynamic friction between the drum and the conveyor belt are $\mu_s = 0.2$ and $\mu_d = 0.1$. The conveyor belt was modelled as a single body (elastic). Due to the comparative nature of the carried out studies, the belt has been modelled as an isotropic material. Mechanical properties of the belt which are used in the numerical research are presented in Table 1.

Table 1. Mechanical properties of the conveyor belt.

Property	Unit	Value
Density	kg/m^3	2000
Young's modulus	MPa	3000
Poisson ratio	–	0.2

The elements through which the conveyor belt was lifted were also treated as non-deformable. The mechanical properties given to the drum and the elements through which the belt was lifted corresponded to steel (Table 2).

Table 2. Mechanical properties of the steel parts.

Property	Unit	Value
Density	kg/m^3	7830
Young's modulus	MPa	207000
Poisson ratio	–	0.3

In the conducted research, it is assumed that the weight of the conveyor belt is 2.5 tons. To reduce the calculation cost, the entire belt was not modeled. The belt has been shortened and loaded at the both ends with an F force with value of 12500 N. This force results from the assumed belt weight. Load increased from 0 to 12500 N in 1.5 s (see Fig. 2).

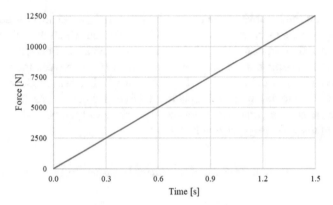

Fig. 2. Graph of load increment as a function of time.

2.1 Conveyor Belt Lifting with a Chain

As it was mentioned earlier, the created model consists of three elements. Conveyor belt and drum were modeled with the use of shell elements. In the case of a conveyor belt, the finite element edge length was 10 mm. The drum was discretized and as a result it was divided into 10 mm edge-length shell elements. In this case belt was lifted with the use of the chain consisting of 23 links. Dimensions of the single chain link are shown in Fig. 3.

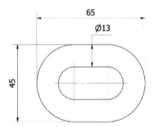

Fig. 3. Dimensions of the chain link.

Geometry of the chain links were discretized (solid elements). The chain links were treated as non-deformable and were given mechanical properties corresponding to steel (see Table 2). However, chain links can move relative to each other.

The contact between the chain links takes into account static and dynamic coefficients of friction ($\mu_s = 0.2$ and $\mu_d = 0.1$). The same friction coefficients were applied between chain links and conveyor belt. The boundary conditions which were applied on the model are the removal of all degrees of freedom from the drum (situation in which it is blocked by the engine). As it was mentioned earlier at both ends of the conveyor belt, force derives from its weight was applied (F = 12500 N). Five degrees of freedom were also taken away from the single node in the last links of the modeled chain (possibility to rotational movement on the symmetry plane was left). The prepared model and applied boundary conditions are presented in Fig. 4.

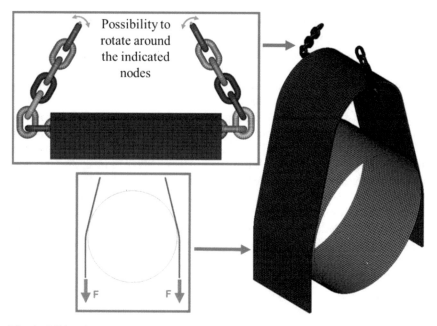

Fig. 4. Lifting the conveyor belt with the use of a chain - model and boundary conditions.

Figure 5 presents a colored map of the Von Mises stresses distribution on the conveyor belt lifted through the chain. As it could be seen, the conveyor belt has undergone significant deformations.

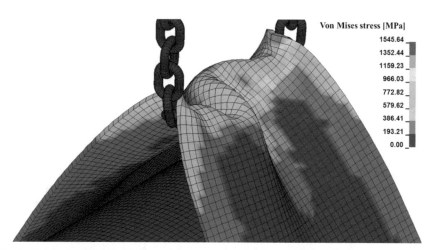

Fig. 5. Colored map of Von Mises stress distribution in case of the conveyor belt lifted with the use of the chain.

The chain links have been strained and the belt has been deformed into the shape of a 'saddle'. The highest Von Mises stress values were observed in the areas where the chain links squeeze the edges of the conveyor belt inwards. The conducted studies indicate that in this area, as a result of belt lifting through the chain, the conveyor belt may be damaged. As a result of such damage, the strength of the conveyor belt may be reduced. In the case of conveyor belts made from composite material, damage of the top layer (made from rubber) results in the exposure of reinforcing fibers. It exposes them to environmental impact and mechanical damage. It is also worth noticing that, the load is transferred through the edge of the belt (higher values of the Von Mises stress at the edge of the belt than in the middle). Von Mises stress values will not be quantified because the belt has been modelled in a simplified way as an isotropic material. Strength analysis of the conveyor belt made from composite material should be performed with use of e.g. Tsai-Wu criterion. Conveyor belt should be also divided into the three layers which are mentioned in introduction. In the next chapter of the article the stress distribution inside conveyor belt which is lifted with the use of various elements (chain and pipe) will be compare.

2.2   Conveyor Belt Lifting with a Pipe

In the second of the analyzed cases, the conveyor belt was lifted by a 50 mm diameter pipe. This element has been treated as non-deformable and has been assigned mechanical properties corresponding to steel (see Table 2). As a result of discretization, geometry of the pipe was divided into shell elements with an edge length of 2 mm. The models of the conveyor belt and drum were made as it was presented in Chapter 2.1. The boundary conditions applied on the model are the removal of all degrees of freedom from the drum (situation in which it is blocked by the engine) and the applied force of F = 12500 N to both ends of the conveyor belt. All degrees of freedom have also been taken away from the pipe through which the belt is lifted. The model prepared in such a way and applied boundary conditions are presented in Fig. 6.

Figure 7 shows a colored map of Von Mises stress distribution in conveyor belt lifted through a steel pipe. As it could be seen in case of conveyor belt being lifted through the pipe, the Von Mises stress distribution is completely different than in case in which the belt is lifted up with use of the chain. When the belt is lifted up with use of the pipe, the stress distribution is more uniform through its whole width. The biggest value of the Von Mises stress is observed in the area in which the belt was pulled up by the pipe. This is a similar situation to the case in which the belt was lifted with use of the chain (the biggest values of Von Mises stress were obtained in the area which is pulled by the lifting element). However, in the case in which the belt is lifted with use of the pipe, the biggest values of Vin Mises stress are much smaller (almost three times) than in case in which belt is lifted with use of the chain.

Deformation of the conveyor belt is also smaller. In this case the conveyor belt is mainly starched at longitudinal direction. It is worth noticing this is one of the direction in which reinforcing fibers are arranged (direction in which strength of the belt is the highest). Based on this assumption, the lifting of the conveyor belt is carried with use of the pipe, which significantly reduces the risk of damage of the belt during its install or dismantle.

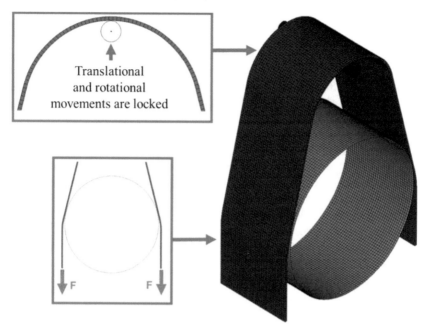

Fig. 6. Lifting the conveyor belt with a pipe - model and boundary conditions.

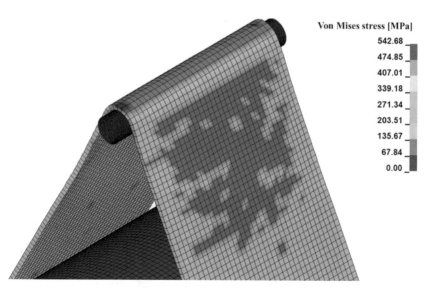

Fig. 7. Colored map of Von Mises stress distribution in case of the conveyor belt being lifted with the use of the pipe.

3   Conclusions

The presented Von Mises stress distribution maps indicate that, the lifting of the conveyor belt with the use of the chain results in stress concentration on its edges. It could lead to the situation in which both the protection material (rubber) and the reinforcing fibers could be damaged. Compression of the belt edge by adjacent chain links can also cut the belt or significantly reduce its strength.

Conveyor belt lifted with the use of the pipe does not cause such large deformations of the belt. It also results in a homogeneous stress distribution field without stress concentration along the edge of the belt. An additional advantage resulting from the absence of excessive deformation of the belt is that it extends in the longitudinal direction (in which reinforcing fibers are arranged), and thus in the direction in which the conveyor belt is characterized by the highest strength. When conveyor belt is lifted with the use of the chain beyond the local concentration of stress, the reinforcing fibers are also deformed. Load is not transferred in the direction in which fibers are arranged. It could result in belt damage under much less load.

In the case of the conveyor belt being lifted with the use of the 50 mm diameter pipe the highest value of Von Mises stress is three times smaller than in the case when belt is lifted with the use of the chain. Deformation, stress distribution and its large increase in the case of the conveyor belt being lifted with use of the chain shows that the assembly and disassembly of the belt should not be performed with the use of the chain. Conveyor belts should be lifted with the use of stiff elements with round geometry.

The results presented in the paper are obtained in preliminary studies. These studies will be continued. Models will be improved and validated with an experimental research.

References

1. Barburski, M.: Analysis of the mechanical properties of conveyor belts on the three main stages of production. J. Ind. Text. **45**(6), 1322–1334 (2016)
2. Sarkar, P.P., Ghosh, S.K., Gupta, B.R., Bhowmick, A.K.: Studies on adhesion between rubber and fabric and rubber and rubber in heat resistant conveyor belt. Int. J. Adhes. Adhes. **9**(1), 26–32 (1989)
3. Hou, Y.F., Meng, Q.R.: Dynamic characteristics of conveyor belts. J. China Univ. Min. Technol. **18**(4), 629–633 (2008)
4. Fedorko, G., Molnar, V., Grincova, A., Dovica, M., Toth, T., Husakova, N., Taraba, V., Kelemen, M.: Failure analysis of irreversible changes in the construction of rubber–textile conveyor belt damaged by sharp-edge material impact. Eng. Fail. Anal. **39**, 135–148 (2014)
5. Fedorko, G., Molnár, V., Živčák, J., Dovica, M., Husáková, N.: Failure analysis of textile rubber conveyor belt damaged by dynamic wear. Eng. Fail. Anal. **28**, 103–114 (2013)
6. Sagami Conveyors CO. https://www.sagami-c.co.jp/service/construction/. Accessed 03 Oct 2020
7. Hebei Global Technology CO., http://www.hbhqrubber.com/Bucket-Elevator-Belt-Conveyor-pd6369046.html. Accessed 03 Oct 2020

The Active Synthesis in Vibration Reduction Using the Example of Driving Systems

Tomasz Dzitkowski$^{(\boxtimes)}$ ⓘ and Andrzej Dymarek ⓘ

Faculty of Mechanical Engineering Institute of Engineering Processes Automation and Integrated Manufacturing Systems, Silesian University of Technology, Konarskiego 18A, Gliwice, Poland
tomasz.dzitkowski@polsl.pl

Abstract. The paper presents application of the active synthesis method for determining parameters of identified drive systems of machines together with parameters of active reduction of vibration. The presented method uses the comparative synthesis method to determine the parameters of vibration reduction. The gist of the method is using dynamic properties, assumed at the beginning of the task, in the form of resonance and anti-resonance frequencies and vibration amplitudes.

Keywords: Dynamic flexibility · Mobility · Accelerance · Natural circular frequencies · Damped circular frequencies

1 Introduction

The sub-system of a machine, which dynamic properties could be solved the same way for a large group of devices, is the drive system. The development of technology has forced from drive systems designers providing high durability and reliability during work period of these devices. Therefore, already in the design stage, one needs to take care of these problems, which can cause such disruption in their work that will result in a considerable deterioration of their work conditions. Designing technical means using only traditional static methods proves to be insufficient. The solution, which seems to guarantee the achievement of the intended designing objectives, is the development of algorithms for modeling, analysis and synthesis of dynamical systems [1–15]. The research on the passive and active synthesis of mechanical systems is developed by the authors of this paper. The scientific problem, taken in works [5–9], concerned the synthesis of machine driving systems as torsional vibration models. Synthesis of torsional vibrating mechanical systems is understood, in these works, as a calculation method by which the mechanical system is designed with parameters that complying the desired characteristic in the form of chosen resonance and anti-resonance frequencies. The synthesized structure of mechanical system consists of non-deformable discs with one degree of freedom connected with each other with weightless sections of shafts. The non-deformable discs model elements mounted on a shaft like: gears, pulleys, flywheels, and also couplings etc. that is elements, whose diameters are larger than the external diameter of the shaft. The components of the non-deformable discs could be also the mass moments of inertia

A. Mężyk et al. (Eds.): SMWM 2020, AISC 1336, pp. 61–72, 2021.
https://doi.org/10.1007/978-3-030-68455-6_6

of the shaft sections located between the discs. The next step presents an active synthesis of the obtained machine drive system models. This task makes it possible to modify the pre-assumed dynamical characteristics in the form of the resonance and anti-resonance frequency sequence, by reducing resonance frequencies to the required values of the amplitude of vibration of the system. The tool that makes possible to modify the characteristics is the determined controlling moment to generate which it is required to provide an additional source of energy to the system.

This work is a continuation of research on development of methods of synthesis of passive and active reduction of vibration in mechanical systems. This study presents the use of active synthesis methods to determine parameters of moments, reducing vibrations of selected resonance frequencies. At the same time, it enables implementation of values of the moments in the system by means of passive damping components, as well as by combination of active and passive components in the system. Such extension of the synthesis task gives the designer a great number of possibilities for selecting optimal parameters of the designed system.

2 Synthesis of Driving Systems

The first step in the synthesis of mechanical systems is to search for mathematical functions which, on the one hand, satisfy the conditions of the designed systems and, on the other hand, can be realized in a real system. The method of determining the analytical form of the dynamic characteristic presented in the work consists in accepting the series of resonance frequencies and the anti-resonance ones (poles and zeros of the desired dynamic characteristic) in the form:

$$\omega_{b1} < \omega_{z1} < \omega_{b2} < \omega_{z2} < \ldots < \omega_{z(i-1)} < \omega_{bi},$$

or

$$\omega_{b1} < \omega_{z1} < \omega_{b2} < \omega_{z2} < \ldots < \omega_{bi} < \omega_{zi}$$

where: $\omega_{b1}, \omega_{b2}, \ldots, \omega_{bi}$ – resonance frequencies (circular frequencies of the free vibration of the undamped system); $\omega_{z1}, \omega_{z2}, \ldots, \omega_{zi}$ – anti-resonance circular frequencies; $i = 1, 2, 3, \ldots, n$. Basing on such dynamic properties, characteristic functions are defined in the form of dynamic flexibility $Y(s)$ or dynamic stiffness $Z(s)$, defined as:

$$Y(s) = \frac{1}{Z(s)} = Y_{11}(s) = \frac{\varphi_1(s)}{F_1(s)} \qquad (1)$$

where: s – complex variable; $\varphi_1(s)$ – Laplace transform of the angular displacement of the first synthesized inertial element, determined at zero initial conditions; $F_1(s)$ – Laplace transform of the inducing moment with respect to the first synthesized inertial element determined at zero initial conditions. In the case when the number of circular resonance frequencies $l.\omega_{bi}$ is larger than the anti-resonance ones $l.\omega_{zi}$. ($l.\omega_{bi} > l.\omega_{zi}$), the characteristic function adopts the following form of dynamic flexibility (dynamic stiffness):

$$Y(s) = \frac{1}{Z(s)} = \frac{\left(s^2 + \omega_{z1}^2\right)\left(s^2 + \omega_{z2}^2\right) \ldots \left(s^2 + \omega_{z(n-1)}^2\right)}{s^2\left(s^2 + \omega_{b2}^2\right)\left(s^2 + \omega_{b3}^2\right) \ldots \left(s^2 + \omega_{bn}^2\right)} \qquad (2)$$

Basing on the determined characteristics, the class of designated mechanical structures can be determined. In the case of the analyzed characteristic (2) in which the resonance frequency is zero ($\omega_{b1} = 0$), these are semi-definite (free) systems. These systems are characterized by the absence of constraints on the displacements and the speeds of the individual components of the system structure. In the case of such determined functions, which describe the dynamic properties of the vibrating discrete system as rigidity or dynamic susceptibility, the following transformations are used:

$$V(s) = sY(s) \text{ or } I(s) = s^2 Y(s) \tag{3}$$

and

$$U(s) = \frac{1}{s} Z(s) \text{ or } M(s) = \frac{1}{s^2} Z(s) \tag{4}$$

where: $V(s)$ – mobility (mechanical admittance); $I(s)$ – inertance; $U(s)$ – slowness (mechanical impedance); $M(s)$ – virtual mass. Dynamic functions, used in the synthesis of mechanical systems, are functions of mobility and their inverse - of slowness. By introducing the synthesis of slowness characteristics into a continued fraction (other methods are described in [2–9]), this function is expressed in the form of the rational function of the complex variable s (quotient of two polynomials):

$$U(s) = \frac{s^l + d_{l-2}s^{l-2} + \cdots + d_1 s}{s^{l-1} + e_{l-3}s^{l-3} + \cdots + e_0} \tag{5}$$

Finally, the slowness (5) is obtained in the form of a continued fraction:

$$U(s) = I_1 s + \cfrac{1}{\frac{s}{c_1} + \cfrac{1}{I_2 s + \cfrac{1}{\frac{s}{c_2} + \frac{1}{I_3 s} \cdots + \cfrac{1}{\frac{s}{c_{n-1}} + \frac{1}{I_n s}}}}} \tag{6}$$

where: $I_i [\text{kgm}^2]$, $i = 1, 2, \ldots, n$ – values of mass inertial moments of synthesized inertial elements of the analyzed drive system; $c_i [\text{Nm/rad}]$, $i = 1, 2, \ldots, n-1$ – values of stiffness coefficients of the synthesized elastic elements of the drive system being analyzed. As the result of the conducted synthesis of slowness (5), one obtains a discrete vibrating system shown in Fig. 1. In the figure is shown the input signal (exciting torque) and the output signal (angular velocity) with respect to which the slowness characteristic (5) was defined.

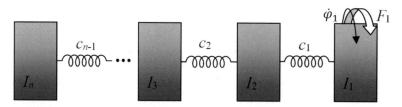

Fig. 1. Cascade structure of a free driving system.

The set of determined values of dynamic parameters for the structure of Fig. 1 should be treated as a "representative" of the possible solutions. The synthesis task allows obtaining an infinite set of these parameters within one assumed structure and its dynamic properties. For this purpose, the dimensionless proportional coefficient H is introduced into the analyzed slowness characteristics. This allows determining an infinite set of solutions of dynamic parameters with respect to the assumed dynamic properties. The introduction of the proportionality coefficient, in the case of the analyzed slowness, in the form of a continued fraction (6) will take the form:

$$HU(s) = HI_1 s + \cfrac{1}{\cfrac{s}{Hc_1} + \cfrac{1}{HI_2 s + \cfrac{1}{\frac{s}{Hc_2} + \frac{1}{HI_3 s} \cdots + \cfrac{1}{\frac{s}{Hc_{n-1}} + \frac{1}{HI_n s}}}}} \tag{7}$$

where: HI_i – set of mass moments of inertia, dependent on the value of the proportionality factor; Hc_{i-1} – set of rigidity of elastic elements, dependent on the value of the proportionality factor; $H \in (0, \infty)$; $i = 1, 2, \ldots, n$.

3 Active Synthesis of Driving Systems

The active synthesis consists in taking into account in the characteristics (2) the circular frequency of free damping vibration ω_{ti}. This is done by introducing into the characteristics the damping coefficients h_i described with the relation:

$$h_i = \frac{\delta_i}{2\pi}\omega_{ti} \tag{8}$$

where:δ_i – logarithmic damping decrement. Then the values, taken into account in the characteristics of free damping vibration frequencies ω_{ti}, could be expressed as a dependency:

$$\omega_{ti} = \sqrt{\omega_{bi}{}^2 - h_{bi}^2} \tag{9}$$

while the dynamic susceptibility takes the form:

$$Y1(s) = \frac{\prod_{i=1}^{n-1}(s^2 + 2h_{zi}s + \omega_{zi}^2)}{s^2 \prod_{i=2}^{n}(s^2 + 2h_{bi}s + \omega_{bi}^2)} \tag{10}$$

The result of introducing the circular frequency of free damped vibration is the modification of the modification of the characteristic, which graphical effect is shown in Fig. 2.

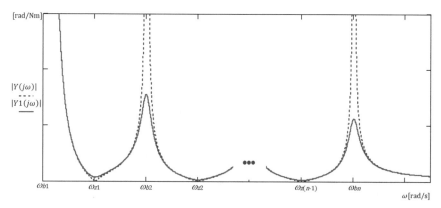

Fig. 2. Characteristics of dynamic flexibility (2) taking into account the vibration damping coefficient.

The values of damping coefficients h_{b1}, h_{b2}, ..., h_{bn}, entered to the dynamic characteristic are determined from the assumed, required values of the amplitude of the forced vibrations. These amplitudes are the maximum angular distortion, angular velocity or angular acceleration of the first inertial element corresponding to the response of the system at the assumed amplitude of the extorting torque. Any damping coefficient h_{bi} is the solution of the following equation system:

- in the case of the desired angular displacement amplitude:

$$\begin{cases} s = j\omega_{bi} \\ |\varphi_1(s, h)| = |F_1(s)Y1(s, h)| \\ |\varphi_1(s, h)| = A_{b\varphi i} \end{cases} \tag{11}$$

- with respect to the desired angular velocity amplitude:

$$\begin{cases} s = j\omega_{bi} \\ |s\varphi_1(s, h)| = |F_1(s)V1(s, h)| \\ |s\varphi_1(s, h)| = A_{b\dot\varphi i} \end{cases} \tag{12}$$

- for the desired angular acceleration amplitude:

$$\begin{cases} s = j\omega_{bi} \\ |s^2\varphi_1(s, h)| = |F_1(s)I1(s, h)| \\ |s^2\varphi_1(s, h)| = A_{b\ddot\varphi i} \end{cases} \tag{13}$$

The number of solutions corresponds to the number of introduced resonance frequency damping coefficients. The graphical interpretation of determining the value h_{bi} taking into account the required amplitude of vibration deflection is presented in Fig. 3.

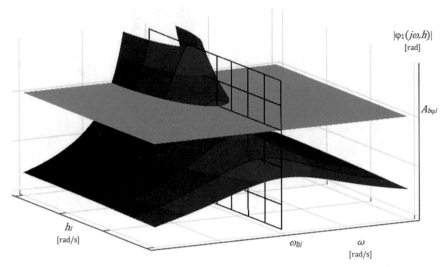

Fig. 3. Graphical interpretation of determining the damping coefficient h_{bi} taking into account the required displacing vibration amplitude.

In the analyzed case of modification, determining only one damping coefficient is required. The values of introduced into the dynamic characteristics the damping coefficients h_i are dependent on each other and are determined by the way in which act the active torques or the passive damping parameters are applied. In the case of active elements located parallel to the elastic elements (the case considered at work), the values of the damping coefficients are calculated by determining the proportionality factor λ:

$$\lambda = 2h_{bi}/\omega_{bi}^2 \qquad (14)$$

where: ω_{bi} – values of resonance frequencies, h_{bi}– value of a damping coefficient in relation to the resonance frequency ω_{bi}, which for the other resonance and anti-resonance frequencies enables determining the damping coefficients as it is presented below:

$$h_{b(z)i} = \lambda\omega_{b(z)i}^2/2 \qquad (15)$$

In order to dampen the vibration of the resonance zones of the drive system, it is assumed acting of the additional dynamic parameters in the form of active moments M_f (located parallel to the elastic elements) what is presented in Fig. 4:

$$\begin{cases} M_{f1}(t) = k_{\dot{\varphi}1}\dot{\varphi}_1(t) \\ M_{f2}(t) = k_{\dot{\varphi}2}(\dot{\varphi}_1(t) - \dot{\varphi}_2(t)) \\ \qquad \vdots \\ M_{fn}(t) = k_{\dot{\varphi}n}(\dot{\varphi}_1(t) - \dot{\varphi}_n(t)) \end{cases} \qquad (16)$$

where: $k_{\dot{\varphi}i}$[Nms/rad] – coefficients of gain of moments of active elements located parallel to the two-unit components of the elastic type, depending on angular velocity $\dot{\boldsymbol{\varphi}}(t) = \begin{bmatrix}\dot{\varphi}_1(t), \dot{\varphi}_2(t), \ldots, \dot{\varphi}_n(t)\end{bmatrix}^T$ corresponding to the generalized coordinates of points determining the location of the inertial elements (see Fig. 4), $i = 1, 2, \ldots, n$.

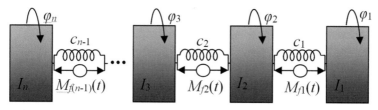

Fig. 4. Cascade structure of a driving system with active vibration reduction components situated parallel to the elastic elements.

Next the momentum gain coefficients are determined. Basing on the synthesized structure and parameters of the drive system, the dynamic stiffness matrix is created:

$$
\mathbf{Z}_s(s) =
\begin{bmatrix}
I_1 s^2 + c_1 & -c_1 & \cdots & 0 \\
-c_1 & I_2 s^2 + c_1 + c_2 & \cdots & 0 \\
\vdots & \vdots & & \vdots \\
0 & 0 & \cdots & I_n s^2 + c_{n-1}
\end{bmatrix}
\tag{17}
$$

and, using the dependency (16), the matrix of the active torque coefficients $\mathbf{M}_f(s)$:

$$
\mathbf{M}_f(s) =
\begin{bmatrix}
k_{\dot\varphi 1} s & -k_{\dot\varphi 1} s & \cdots & 0 \\
-k_{\dot\varphi 1} s & k_{\dot\varphi 1} s + k_{\dot\varphi 2} s & \cdots & 0 \\
\vdots & \vdots & & \vdots \\
0 & 0 & \cdots & k_{\dot\varphi n} s
\end{bmatrix}
\tag{18}
$$

Basing on the $\mathbf{M}_f(s)$ matrix and on the stiffness matrix $\mathbf{Z}(s)$ one could determine the following polynomial:

$$
\det\big(\mathbf{Z}_s(s) + \mathbf{M}_f(s)\big) = A_{2n} s^{2n} + A_{2n-2} s^{2n-2} + A_{2n-4} s^{2n-4} + \cdots + A_2 s^2
\tag{19}
$$

In order to calculate the values of active moments, the resulting polynomial must be divided by the A_{2n} coefficient and then equalized to the polynomial characterizing the assumed dynamic properties in the form of resonance frequencies and coefficients of free vibration damping. These equations take the following form:

$$
\frac{\det\big(\mathbf{Z}_s(s) + \mathbf{M}_f(s)\big)}{A_{2n}} = s^2 \prod_{i=2}^{n}\big(s^2 + 2h_{bi} + \omega_{bi}{}^2\big)
\tag{20}
$$

After comparing the coefficients at the same powers of polynomials of Eq. (20), basing on the created system of equations, the parameters of active torques reducing vibration of the identified system are determined.

4 Numerical Example

In this chapter is presented a numerical example of determining parameters and a structure of a drive system fulfilling dynamical properties, assumed in advance, in the form:

- resonance frequencies: $\omega_{b1} = 0 \, \text{rad/s}$, $\omega_{b2} = 150 \, \text{rad/s}$, $\omega_{b3} = 300 \, \text{rad/s}$,
- anti-resonance circular frequencies: $\omega_{z1} = 80 \, \text{rad/s}$, $\omega_{z2} = 240 \, \text{rad/s}$,
- amplitudes: $A_{b\varphi 3} = 0.004 \, \text{rad}$,
- proportionality factor: $H = 1$.

In the first step of the synthesis is created a dynamic characteristic basing on the accepted resonance frequencies, in the form of poles and zeros, in accordance with Eq. (3):

$$Y(s) = \frac{1}{Z(s)} = \frac{(s^2 + 80^2)(s^2 + 240^2)}{s^2(s^2 + 150^2)(s^2 + 300^2)} \tag{21}$$

On the basis of Eq. (4) the resulting characteristic Eq. (21) is modified to the slowness function, and using Eq. (5) the parameters (see Table 1) and the structure of the system are determined (Fig. 5).

Table 1. Value of elastic and inertial components.

I, kgm^2	c, Nm/rad
$I_1 = 1$	$c_1 = 48500$
$I_2 = 1{,}625$	$c_2 = 35424.593$
$I_3 = 2{,}868$	

The next step in identification of the system characteristic, after determining the parameters of the mechanical system, is modification the Eq. (21) by introducing the parameters of quenching in resonance frequencies of the system:

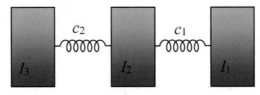

Fig. 5. Torsional system obtained by decomposition of the slowness function.

$$Y1(s) = \frac{(s^2 + 2h_{z1}s + 80^2)(s^2 + 2h_{z2}s + 240^2)}{s^2(s^2 + 2h_{b2}s + 150^2)(s^2 + 2h_{b3}s + 300^2)} \tag{22}$$

where h_{b2}, h_{b3} the values of resonance frequency damping coefficients corresponds to the values $\omega_{b2}, \omega_{b3}, h_{z1}, h_{z2}$ the values of anti-resonance frequency damping coefficients corresponds to the values ω_{z1}, ω_{z2}. Basing on the assumed vibration amplitudes, using the Eq. (11), the value of h_{b3} have been determined. In the analyzed case it takes the following value: $hb3 = 1.8586$ rad/s. The next values of the damping coefficients are calculated by determining the proportionality factor λ:

$$\lambda = 2h_{b3}/\omega_{b3}^2 = 0.0000413 \tag{23}$$

which for the other resonance and anti-resonance frequencies enables determining the damping coefficients as it is presented below:

$$\begin{cases} h_{b2} = \lambda\omega_{b3}^2/2 = 0.4646\,\text{rad/s} \\ h_{z1} = \lambda\omega_{z1}^2/2 = 0.1322\,\text{rad/s} \\ h_{z2} = \lambda\omega_{z2}^2/2 = 1.1894\,\text{rad/s} \end{cases} \tag{24}$$

The graphical representation of the dynamic characteristic of the analyzed system, before and after modification, is shown in Fig. 6.

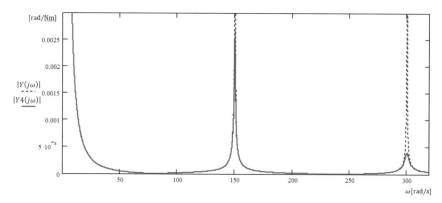

Fig. 6. Dynamic characteristic of the system subjected to synthesis.

To obtain the determined vibration amplitude of resonance frequencies of the analyzed system (Fig. 5), it is assumed acting of additional moments in the system (Fig. 7), in accordance with Eq. (16).

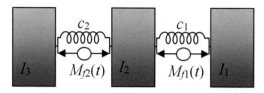

Fig. 7. Torsional system with active components of vibration reduction.

Basing on the synthesized structure and parameters of the drive system, the dynamic stiffness matrix is created:

$$\mathbf{Z}_s(s) = \begin{bmatrix} 1s^2 + 48500 & -48500 & 0 \\ -48500 & 1.625s^2 + 83924.593 & -35424.593 \\ 0 & -35424.593 & 2.868s^2 + 35424.593 \end{bmatrix} \tag{25}$$

and the matrix of the active torque coefficients $M_f(s)$:

$$M_f(s) = \begin{bmatrix} k_{\dot\varphi 1}s & -k_{\dot\varphi 1}s & 0 \\ -k_{\dot\varphi 1}s & k_{\dot\varphi 1}s + k_{\dot\varphi 2}s & -k_{\dot\varphi 2}s \\ 0 & -k_{\dot\varphi 2}s & k_{\dot\varphi 2}s \end{bmatrix} \tag{26}$$

Basing on the $M_f(s)$ matrix and on the stiffness matrix $Z(s)$ one could determine the following polynomial:

$$\det(Z_s(s) + M_f(s)) = 4.6605s^6 + (7.5285 \cdot k_{\dot\varphi 1} + 4.943 \cdot k_{\dot\varphi 2})s^5 + (524294.9463 + 5.493 \cdot k_{\dot\varphi 1} \cdot k_{\dot\varphi 2})s^4$$
$$+ (194587.2893 \cdot k_{\dot\varphi 1} + 266410.5 \cdot k_{\dot\varphi 2})s^3 + (9.43748 \cdot 10^9 - 1.45519 \cdot 10^{-11} \cdot k_{\dot\varphi 1} \cdot k_{\dot\varphi 2})s^2$$
$$+ (-2.38419 \cdot 10^{-7} \cdot k_{\dot\varphi 1} - 4.76837 \cdot 10^{-7} \cdot k_{\dot\varphi 2})s + (-0.0234) \tag{27}$$

The values of control moments coefficients are determined using the relationship:

$$\frac{\det(Z_s(s) + M_f(s))}{4.6605} = s^2(s^2 + 0.9292s + 150^2)(s^2 + 3.718s + 300^2) \tag{28}$$

In the presented case they are as follows:

$$\begin{cases} k_{\dot\varphi 1} = 2.003 \text{ Nms/rad} \\ k_{\dot\varphi 2} = 1.463 \text{ Nms/rad} \end{cases} \tag{29}$$

The defined values of active moments, due to their configurations in the system, could also be obtained through both passive, damping components (Fig. 8) or by hybrid structure of active-passive components.

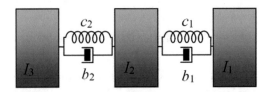

Fig. 8. Torsional system with passive components of vibration reduction.

Parameters of the passive system, take the following form:

$$\begin{cases} b_1 = k_{\dot\varphi 1} = 2.003 \text{ Nms/rad} \\ b_2 = k_{\dot\varphi 2} = 1.463 \text{ Nms/rad} \end{cases} \tag{30}$$

The determined parameters of active vibration reduction allow to reduce vibration to the desired vibration amplitude of selected resonance frequencies. During the simulation, the periodic moment with a unit amplitude and circular frequency corresponding to the resonance frequency of free vibration of the system was assumed as the forcing signal. In addition, it was assumed that in the case of active vibration reduction, the control moments are activated after 1.5 s (Fig. 9).

The generated time characteristics of the reduced systems, with respect to the pre-determined amplitude, confirm the correctness of the synthesis method in reducing vibration due to dynamic properties of a mechanical system.

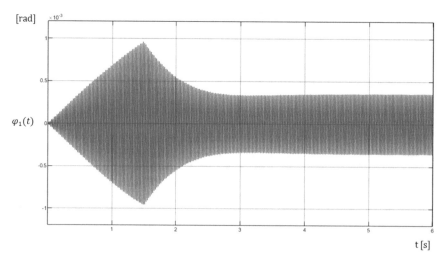

Fig. 9. Rotation angle of the first inertial component, taking into account active reduction.

5 Conclusions

The paper presented discusses the formulation and formalization of the problem of active synthesis of vibrating systems which is to be used as a computer-aided design tool for all operating conditions of the device. The proposed method of active synthesis of discrete mechanical systems is a combination of the passive synthesis methods with the method of determining the active moments. The advantage of the proposed method is that, as a result of the synthesis, one obtains a large number of structures, system parameters and amplifications of control moments which can substantially affect the optimal choice of parameters for the object in question.

References

1. Byrtus, M., Zeman, V.: On modeling and vibration of gear drives influenced by nonlinear couplings. Mech. Mach. Theory **46**(3), 375–397 (2011)
2. Dymarek A., Dzitkowski T., Herbuś K., Ociepka P., Sękala A.: Use of active synthesis in vibration reduction using an example of a four-storey building. J. Vibr. Control January 13 (2020) https://doi.org/10.1177/1077546319898970
3. Dymarek A., Dzitkowski T.: Passive reduction in vibration considering the desired amplitude in the case of damping proportional to the mass. Mechatron. Syst. Mater. VI. Solid State Phenomena **220–221**, 43–48 (2015).
4. Dymarek A., Dzitkowski T.: Inverse task of vibration active reduction of Mechanical Systems. Mathematical Problems in Engineering (2016) Article ID 3191807
5. Dzitkowski, T., Dymarek, A.: Active synthesis of machine drive systems using a comparative method. J. Vibroengineering **14**(2), 528–533 (2012)
6. Dzitkowski, T., Dymarek, A.: Active synthesis of machine drive systems. Appl. Mech. Mater. **430**, 178–183 (2013)
7. Dzitkowski, T., Dymarek, A.: Active reduction of identified machine drive system vibrations in the form of multi-stage gear units. Mechanika **20**(2), 183–189 (2014)

8. Dzitkowski, T., Dymarek, A.: Passive reduction in the identified vibrations of the machine drive system in the form of multistage gear units. Solid State Phenom. **220–221**, 182–187 (2015)
9. Dzitkowski, T., Dymarek, A.: Method of active and passive vibration reduction of synthesized bifurcated drive systems of machines to the required values of amplitudes. J. Vibroengineering **17**(4), 1578–1592 (2015)
10. Park, J.S., Kim, J.S.: Dynamic system synthesis in term of bond graph prototypes. KSME Int. J. **12**(3), 429–440 (1998)
11. Rao, J.S.: History of Rotating Machinery Dynamics. Springer, Dordrecht (2011)
12. Rao, Z., Zhou, C.Y., Deng, Z.H., Fu, M.Y.: Nonlinear torsional instabilities in two-stage gear systems with flexible shafts. Int. J. Mech. Sci. **82**, 60–66 (2014)
13. Redfield, R.C., Krishnan, S.: Dynamic system synthesis with a bond graph approach. Part I: Synthesis of one-port impedances. J. Dyn. Syst. Meas. Control **115**(3), 357–363 (1993)
14. Świtoński, E., Mężyk, A.: Selection of optimum dynamic features for mechatronic drive systems. Automation and Constructions. **17**(3), 251–256 (2008)
15. Wu, J.S., Chen, D.W.: Torsional vibration analysis of gear-branched systems by finite element method. J. Sound Vib. **240**(1), 159–182 (2001)

Creating an Integrated Model of a Technical System with Use of the Mechatronic Features

Krzysztof Herbuś$^{(\boxtimes)}$ (ID)

Faculty of Mechanical Engineering Department of Engineering Processes Automation and Integrated Manufacturing Systems, Silesian University of Technology, Konarskiego 18A, 44-100 Gliwice, Poland
krzysztof.herbus@polsl.pl

Abstract. In the work the description of the creation way of an integration of the virtual model with the virtual controller was presented. In order to integrate the control system with the executive subsystem, a method of defining mechatronic features (MOE) has been proposed, which refer to the drive along with its operation logic as a set of objects, consisting of objects such as: functional features (FOE), coupling relations (RSF) between FOE, sensory features (SEOE), feedback relations (RSFSE) between FOE and SEOE, signal features SOE, a transformation relation (RPM) associated with a given MOE. The work describes the process of creating the integrated model in relation to the Stewart platform, which is the main executive element of the car driving simulator for people with mobility dysfunctions. The 3D model of the considered system was created in the MOE description in the Siemens NX PLM software environment. The virtual controller was prepared with use of B&R Automation Studio software. The OPC server is the integrator that allows to simulate the mechatronic function of the system. In the considered case, the OPC server enables the exchange of an information between the 3D model in the MOE description and the virtual controller using SOE. The developed way of integrating the virtual machine model with the virtual controller allows to minimize the risk of malfunctioning of the real system. The presented methodology allows creating the integrated model in the control context, using various programming environments, however, both the geometrical form of the model and the control algorithm can be changed constantly.

Keywords: Mechatronic features · Integrated model · OPC server

1 Introduction

One of the necessary stages of the computer-aided design and a construction process is the modeling and simulation of the operation of the designed systems. Computer-aided modeling of such systems can be carried out using a different class of systems, enabling the modeling process to be oriented depending on the field of the problem being under consideration [1, 2, 8, 10]. The operation simulation of technical systems can also be carried out in many ways, depending on the purpose of the virtual test. In the context of the rapid development of modern industry, it is necessary to acquire digital data and their processing as one of the stages of creating a virtual twin of a technical system [3, 6, 7, 9]. The idea behind the research was to propose an integrated design approach in the control context, in which the model of the considered system is associated with the control system [4, 5]. Then, one coherent model of the system is created, including the mapping of the executive subsystem (mechanical, pneumatic or electric) and the control subsystem. In the such understood, the integrated model of the technical system, any introduced changes in the model and control algorithm have a direct impact on each other. The proposed method of the integrated design in the context of control can be presented in a generalized form as in Fig. 1. The presented design method assumes the possibility of integrating virtual models of technical systems, created in any type of software, with a virtual or real control system using an integrator. The integrator's task is to provide information about the state of the control object (technical system) to the control system, and to provide information about the operation of the technical system to its virtual model.

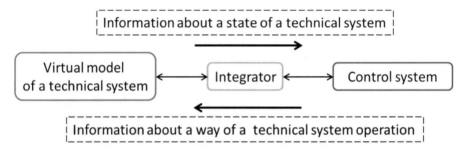

Fig. 1. Generalized scheme of integrated design in the control context.

In order to ensure the possibility of a data exchange between the virtual model of the technical system and the integrator it is necessary to determine the structure of the model. It is also necessary to analyze the possibilities of utilization various integrators and control systems. Due to the problem area under consideration in the structure of the technical system model, one can distinguish among others: objects describing its geometrical form and its system of dimensions, objects describing relations between them and objects describing the state of the system. In order to exchange information between the model and the control system, it is necessary to define additionally signal objects that will be the carrier of information on the state of the object and will be able to transfer information from the integrator concerning the desired operation of the system. The elements that determine the operation of technical systems are drives. Therefore, was proposed that the subsystem directly related to the control system could be isolated from the model of the technical system, and more closely the executive system of a technical device, consisting of drives, signal elements, control and setting ones. Such an isolated subsystem should be integrated with the control system. This approach makes it possible to extract from the model of a technical system, the executive subsystem, and the subsystem describing the geometrical form of the elements affected by the executive subsystem. Therefore, the structure of the model of a technical system should take into account both connections between individual elements of the executive system as well as between them and other elements of the system. In the conducted research was proposed that the design process, in the context of control, should be carried out according to the model structure presented in Fig. 2. Within the model of a technical system, was extracted the structure of the executive subsystem (SPW), the structure of the subsystem describing the geometric form (SPG) of the elements affected by the executive subsystem, technological object and sensory system (USE), which is not directly related to the executive subsystem.

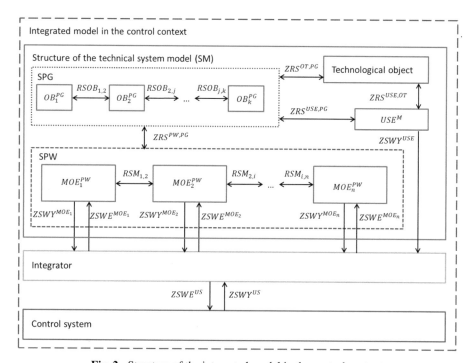

Fig. 2. Structure of the integrated model in the control context.

The flow of information between the technical system model and the integrator is carried out by means of sets of output signals $\text{ZSWY}^{\text{MOE}_n}$, ZSWY^{USE} and input $\text{ZSWE}^{\text{MOE}_n}$. In the presented structure the objects of the type OB_k^{PG} correspond to elements of the technical system that are not part of the executive subsystem, elements of the type $\text{RSOB}_{j,k}$ correspond to the method of linking the objects of the type OB_k^{PG}. On the other hand, the objects of the type ZRS correspond to the coupling relations that occur between the main subsystems of the model of a technical system. Objects of the type ZSWE^{US} and ZSWY^{US} allow the transfer of information between the integrator and the control system.

2 Definition of the Mechatronic Features

In order to integrate the control system with the executive subsystem, was proposed a method of defining mechatronic features (MOE), which refer to the drive along with its operation logic, as a set of objects, consisting of (Fig. 3):

- functional features (FOE), which represent the geometric form of MOE components, and perform a specific function in its area,
- coupling relationships (RSF) between FOE, which determine their behavior,
- sensory features (SEOE), which determine the MOE status by registering the FOE position,
- coupling relations (RSFSE) between FOE and SEOE, which result from the way of linking between SEOE and FOE,
- signal features SOE, which on the basis of RSFSE provide information about the MOE status to the integrator,
- transformation relation RPM associated with the given MOE,
- coupling relations between MOE (RSMF),
- coupling relations between MOE and the geometric form of the subsystem PG (RSMPG).

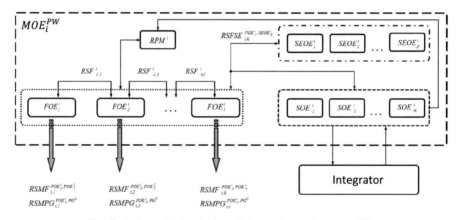

Fig. 3. Schematic description of the mechatronic feature [5].

Mechatronic features can directly cooperate with each other through coupling relations between functional features (RSMF) or with the geometric form of the subsystem PG through the RSMPG coupling relations. However, the main objects characterizing the work of a given MOE are: coupling relations (RSF), the transformation relation (RPM) and information from SEOE features. This approach allows to create mechatronic features databases and their implementation into a model of any technical system, while maintaining their functionality. Which should significantly accelerate the work associated with creating the integrated model of technical means in the control context.

The definition of MOE as a set of FOE, RSF, SEOE, RSFSE, SOE, RSMF, RSMPG and RPM presents the dependence 1 [5].

$$
MOE_i^{PW} = \left\{
\begin{array}{l}
\left(FOE_1^i, FOE_2^i, \ldots, FOE_l^i \right) \\[4pt]
\left(RSF_{1,2}^i, RSF_{2,3}^i, \ldots, RSF_{n,1}^i \right) \\[4pt]
\left(SEOE_1^i, SEOE_2^i, \ldots, SEOE_p^i \right) \\[4pt]
\left(RSFSE_{i,1}^{FOE_1^i, SEOE_1^i}, RSFSE_{i,2}^{FOE_2^i, SEOE_2^i}, \ldots, RSFSE_{i,t}^{FOE_j^i, SEOE_p^i} \right) \\[4pt]
\left(SOE_1^i, SOE_2^i, \ldots, SOE_m^i \right) \\[4pt]
\left(RSMF_{i,1}^{FOE_1^i, FOE_1^i}, RSMF_{i,2}^{FOE_2^i, FOE_3^2}, \ldots, RSMF_{i,k}^{FOE_j^i, FOE_w^k} \right) \\[4pt]
\left(RSMPG_{i,1}^{FOE_1^i, PG^U}, RSMPG_{i,2}^{FOE_2^i, PG^U}, \ldots, RSMPG_{i,r}^{FOE_j^i, PG^U} \right) \\[4pt]
RPM^i
\end{array}
\right\}
\tag{1}
$$

where:

MOE_i^{PW} – mechatronics feature i included in the control sub-system PW,
FOE_l^i – functional feature l included in the mechatronics feature i,
$RSF_{n,l}^i$ – coupling relation between functional features n and l, included in the mechatronics feature i,
$SEOE_p^i$ – sensory feature p included in the mechatronics feature i,
$RSFSE_{i,t}^{FOE_j^i, SEOE_p^i}$ – coupling relation within the mechatronic feature i with the number t, in relation to the functional feature FOE_j^i and sensory feature $SEOE_p^i$,
SOE_m^i – signal feature m included in the mechatronics feature i,
$RSMF_{i,k}^{FOE_j^i, FOE_w^k}$ – coupling relation between mechatronic features i and k, in relation to the functional features FOE_j^i and FOE_w^k,
$RSMPG_{i,r}^{FOE_j^i, PG^U}$ – coupling relation between the mechatronic feature i and the geometric form of the entire system U with the number r, in relation to the functional feature FOE_j^i and the geometric form of the entire system PG^U,
RPM^i – transformation relation associated with the mechatronic feature i,
$i, i, j, k, l, m, n, p, r, t, u, w = 1, 2, \ldots, s; s \in N$.

The presented methodology enables to create the integrated model in the control context, using various programming environments. The integrated model, in the context of control, can be used, depending on the class of applied computer systems, to verify and identify system operation errors, to determine the system workspace, to visualize the operation process in the design phase, for the virtual system startup and for the program verification for the PLC controller.

3   Object of the Integration Process

Creating the integrated model of the technical system in the control context was presented on the example of the car driving simulator. The mentioned test object was chosen because its test minimizes the negative effects of errors in the area of its control algorithm in relation to the real object. In the area of the designed simulator, the stationary hexapod (Stewart's platform) is the main executive system of the driving simulator for people with lower limb dysfunctions (Fig. 4). On the Stewart platform is mounted the body of a Fiat Panda, which has been modified in terms of mechanical equipment. The described system is connected to the control cabinet, which is responsible for the correct operation of the Stewart platform and cooperation with the virtual environment in which the car avatar is moving. The information from the virtual environment and from the modified vehicle's control elements is transmitted, as input signals, to the control cabinet, in which the decision on how the Stewart platform should be controlled is made on the basis of acquired information from the mathematical model.

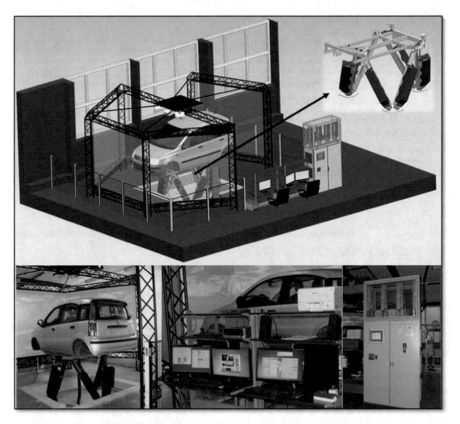

Fig. 4. The Virtual model and the real driving simulator system for people with disabilities [4].

The way of driving the modified vehicle does not differ from the way of driving a car moving in real traffic. This is due to the fact that all control and steering elements from the driver's point of view operate in the same way as in an unmodified car. The whole simulator completes the virtual world projection system.

4   Model of the Drive in the Mechatronic Features Description

In the first step in the area of the hexapod system, the executive subsystem consisting of drives in combination with the drive transmission system was separated. In the presented system there are six the same drives, which were described using the mechatronic features method. A way of describing the single mechatronic feature with reference to the "leg" of the Stewart platform was shown in the Fig. 5. The created mechatronic feature of the actuator consists of the set of functional features (ZFOE) like: drive_shaft, drive_housing, drive_support, nut, bolt, the set of coupling relations between functional features (ZRSMF) like: hinge joint, sliding joint, screw joint, ball joint or mechanical cam, the set of sensory features (ZSEOE) like: distance sensor or speed sensor, the set of signal features which transforming the information about the ZSEOE state to the integrator, and the transformation relation (RPM) (speed control) which get the information from the integrator to the mechatronic feature.

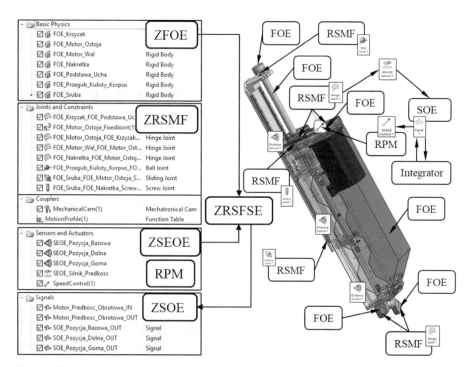

Fig. 5. The actuator model prepared for the integration using the mechatronic features method.

The characteristic feature of the created actuator in the mechatronic features description is that, it can be used many times in the same or another model. In order to create the integrated model of the hexapod, the same model of the single actuator (the mechatronic feature in the "leg" form of the hexapod) was implemented six times in the platform Stewart model. In this way a set of mechatronic features was created. In the next step, the constraints between the mechatronic features (MOE) and the geometric form of the

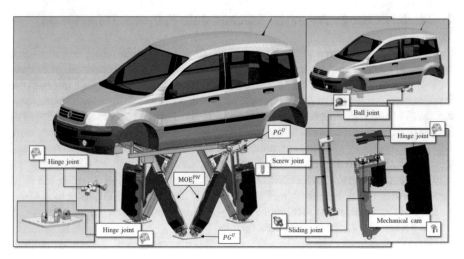

Fig. 6. Hexapod model prepared for simulation of operation in PLM Siemens NX software [cf. 4].

entire modeled (the support, the frame with the car body) system (PG^U) were created, what was shown in the Fig. 6. The set of the transformation relation (RPM) associated with the mechatronic features, determine the behavior of whole hexapod system.

5 The Integrated Model of the Technical System with Use of the Mechatronic Features

The procedure of the preparation of the virtual controller in the B&R Automation Studio software to the integration can be described in the fallowing steps:

- defining variables through which the virtual controller obtains the information about the state of the 3D model of the system,
- defining variables through which the information is transferred to the 3D model of the system about its behavior,
- defining the parameters of the OPC server, which enables the exchange of the information between the 3D model of the system and the virtual controller,
- defining the method of sharing the values of selected variables in the OPC server area (read, write, read/write),
- an implementation of the initial control algorithm on the virtual controller taking into account the state of the 3D model of the analyzed system.

It should be noted that the virtual driver can be created based on the "Standard PC" function, which means that the program runs the virtual driver on the local computer with the assigned IP address 127.0.0.1. In this case, any control algorithm can be tested, but without linking to the actual architecture of the control system. The virtual controller can also be created based on the given hardware architecture of the predicted PLC controller, and individual variables can be assigned to specific controller I/O ports. In this case, both the control algorithm and the controller architecture are tested due to ensuring the correct control process in relation to the consideration technical system.

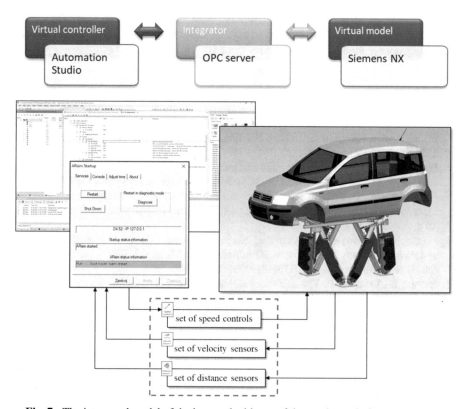

Fig. 7. The integrated model of the hexapod with use of the mechatronic features method.

The created virtual controller was connected via the integrator (OPC server) with the virtual 3D model of the technical system, thus creating the integrated hexapod model in the context of control (Fig. 7). In this case, information about the position of the Stewart platform model are transmitted via the set of distance sensors and set of velocity sensors to the virtual controller, where on the basis of this data and the adopted control algorithm, values in relation to set of speed controls are generated, which describing the realization of the assumed behavior of the hexapod system.

6 Conclusions

The presented methodology of creating integrated models of systems in the context of control is general and can be used to model various types of technical systems. It is also an important tool facilitating the process of designing technical means due to the problem area being considered and the goal of conducting the virtual experiment. The presented way to integrate the model of the executive subsystem with the model of the control system are also general in nature and could be used to implement the task related to the exchange of information between various applications. Based on the created integrated model, it is possible to the safely test various algorithms for controlling the hexapod system in relation to the driver's health and the condition of the tested technical system (the virtual model cannot be damaged or destroyed). And then, the tested algorithm can be safely implemented to the real control system.

References

1. Banas, W., Gwiazda, A., Sekala, A., Foit, K., Cwikla, G.: Positioning a robot in a robotic cell in Tecnomatix. IOP Conf. Ser. Mater. Sci. Eng. **400**, 052002 (2018). https://doi.org/10.1088/1757-899X/400/5/052002
2. Banaś, W., Lysek, K.: Modelling industrial robot fanuc ARC mate 100iB in LabVIEW. Adv. Intell. Syst. Comput. **934**, 1–9 (2019). https://doi.org/10.1007/978-3-030-15857-6_1
3. Grudzinski, M., Okarma, K., Pajor, M., et al.: Visualization of the Workpieces on the CNC Machines Using the Virtual Camera Based on the Sub-pixel IBR Method. Solid State Phenom. **199**, 253–258 (2013)
4. Herbuś, K., Ociepka, P.: Analysis of the Hexapod work space using integration of a CAD/CAE system and the LabVIEW software. IOP Conf. Series: Mater. Sci. Eng. 95, art. no. 012096, (2015) https://doi.org/10.1088/1757-899X/95/1/012096
5. Herbus, K., Ociepka, P.: Virtual commissioning of a robotized production cell with use of mechatronic features. IOP Conf. Ser. Mater. Sci. Eng. **400**, 042030 (2018). https://doi.org/10.1088/1757-899X/400/4/042030
6. Miadlicki, K., Pajor, M., Sakow, M.: Real-time ground filtration method for a loader crane environment monitoring system using sparse LIDAR data. In: 2017 IEEE International Conference On Innovations In Intelligent Systems and Applications (INISTA), pp. 207–212 (2017)
7. Miadlicki, K., Pajor, M., Sakow, M.: Ground plane estimation from sparse LIDAR data for loader crane sensor fusion system. In: 22nd International Conference on Methods and Models in Automation and Robotics (MMAR), pp. 717–722 (2017)
8. Montazeri-Gh, M., Rasti, A.: Comparison of model predictive controller and optimized min-max algorithm for turbofan engine fuel control. J. Mech. Sci. Technol. **33**(11), 5483–5498 (2019). https://doi.org/10.1016/j.precisioneng.2019.08.010
9. Sakow, M., Parus, A., Pajor, M., et al.: Unilateral Hydraulic Telemanipulation System for Operation in Machining Work Area. In: Hamrol, A., Ciszak, O., Legutko, S., Jurczyk, M., (eds.) Advances in Manufacturing (MANUFACTURING 2017), Lecture Notes in Mechanical Engineering, pp. 415–425. Springer, Cham (2018)
10. Zhang, Z., Dong, Y., Han, Y.: Dynamic and control of electric vehicle in regenerative braking for driving safety and energy conservation. J. Vib. Eng. Technol. **8**(1), 179–197 (2020). https://doi.org/10.1007/s42417-019-00098-0

Analysis of Off-Grid Hybrid Solar – Wind Power Plant

Mariola Jureczko[✉] [iD]

Department of Theoretical and Applied Mechanics, Faculty of Mechanical Engineering, Silesian University of Technology, 18A Konarskiego Street, 44-100 Gliwice, Poland
Mariola.Jureczko@polsl.pl

Abstract. In this paper was presented and discussed a numerical model of a hybrid wind-solar power plant was created in Matlab®/Simulink software. A parallel connection between two potential renewable energy sources has been proposed: photovoltaic panels (PVs) and the wind turbine. For numerical simulations, environmental data were used with wind and solar data for Katowice, Silesia City.

One of the main requirements for the created numerical model of a hybrid power plant was to ensure the possibility of changing the most important simulation parameters. Therefore, in the photovoltaic part of the numerical model, it is possible to change the amount of PV panels, their efficiency, and surface area. As for the wind turbine part of the numerical model, it is possible to alter some of the parameters, e.g. swept area, the minimum (cut-in) wind speed, and maximum (cut-out) wind speed. The developed numerical model also includes batteries for which the value of their capacity can be changed. In this model, it is also possible to modify the value of the average daily power demand of the designed installation.

The developed numerical model of the hybrid power plant can be used for preliminary estimation and selection of basic components of a hybrid wind-solar power plant.

Keywords: Wind energy · Wind turbine · Solar energy · Photovoltaic panels · Hybrid power generation system

1 Introduction

For several years, renewable energy sources have been of interest to both scientists and producers. The reasons for this interest are that thanks to the legislation of developed countries and the depletion of conventional energy sources, renewables will play a major role in energy in the future. Also, the continuous expansive use of non-renewable energy sources by man causes global climate change and increasing pollution of the Earth's atmosphere. The use of renewable energy sources by installing micro-installations by individual consumers (single-family houses, small farms, small businesses) is one of the ways to reduce the effects of over-exploitation of fossil deposits. On the other hand, micro-installations using renewable energy sources, due to their specificity, e.g. dependence on the time of day and year, weather or geographic location, cannot be the main source of energy in the current state of technology development.

© The Author(s), under exclusive license to Springer Nature Switzerland AG 2021
A. Mężyk et al. (Eds.): SMWM 2020, AISC 1336, pp. 83–94, 2021.
https://doi.org/10.1007/978-3-030-68455-6_8

An alternative idea is to use more efficient hybrid solutions, i.e. based on at least two renewable energy sources. Solar-wind micro-installation is the most frequently used hybrid system in unconventional power plants. This is because these two renewable energies complement each other perfectly.

In Poland, wind turbines achieve the highest efficiency in late autumn, winter, and the beginning of spring. Photovoltaic panels or solar panels, on the other hand, produce the most energy in summer. The use of a hybrid power plant enables the production of energy even if one of the members of the power plant is inefficient during a given period of the year. The performance of this hybrid system increases when the sun is shining, and the wind is blowing at the same time.

2   Design Assumptions

After consideration of many factors, it was decided that the considered the power plant would be a domestic, the hybrid solar-wind power plant, with the possibility of energy storage in the external battery provided with a current control system. It will aim of supporting the supply of a 200 m^2 single-family house in electricity. The house is in the temperate climate zone, in Poland in the Silesian voivodeship. A general diagram of off-grid the hybrid solar – wind power plant is presented in Fig. 1.

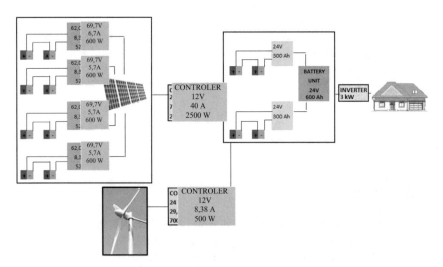

Fig. 1. A general diagram of off-grid the hybrid solar – wind power plant.

Based on the data from the calculator provided by the electricity supplier in Poland and the data from an exemplary household, it can be assumed that the average annual electricity demand is 750 kWh [1]. This demand is variable and strongly depends on the prevailing season, weather, and cloudiness. An example of the amount of energy required for an average family of four living in a 200 m^2 single-family house is based on data from 1 year, is shown in Fig. 2.

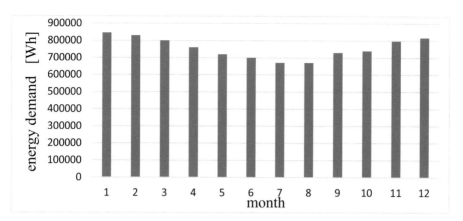

Fig. 2. Chart of the amount of energy required in Wh per month.

The wind level in the area in question in the years 2010–2020 has been averaged each day and presented as an example of the wind level during the year, as shown in Fig. 3, was estimated on the basis of data from the Polish Institute of Meteorology and Water Management – National Research, meteorological station in Katowice [2]. These data were used to perform analyses in the MATLAB®/Simulink environment.

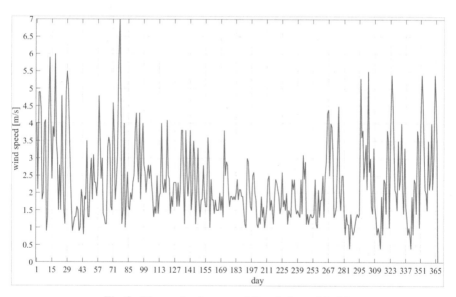

Fig. 3. The graph of average daily wind speed [m/s].

The power contained in the air stream can be determined from the following:

$$P = E \cdot \Delta\vartheta = \frac{1}{2} \cdot \rho \cdot A \cdot V_{inf}^3 \tag{1}$$

where:

ρ – air density,
E – kinetic energy of air movement,
V_{inf} – infinity wind speed,
A – swept area,
$\Delta\vartheta$ – increase in the volume of incoming air.

The mechanical power of the wind turbine is determined by the following:

$$P_{WT} = P \cdot C_p, \tag{2}$$

where:

C_p – Betz coefficient, assumed equals 0.4.

Total radiation is the sum of direct and diffuse radiation on the falling surface. Irradiance, while is the instantaneous power density value of solar radiation perpendicular to the radiation direction of a 1 m^2 surface in one second. Irradiation, on the other hand, is the sum of the intensity of solar radiation counted at a given time (most often a month).

The sunshine in Poland on an annual basis is very uneven; between April and September accounts for 80% of total annual sunshine. In the central part of the Silesian voivodeship, the sun is the smallest, below 1000 kWh/m^2 per year.

To take full advantage of solar energy, you need to determine the optimal angle for setting the photovoltaic panels. Assuming that the panels are permanently mounted, precisely in the direction of solar radiation, it can be assumed that the panels should be mounted at an angle approximate to the latitude at which they will be located and facing south [3, 4]. For Katowice, this is a width of ~50°. According to the Satellite Application Facility on Climate Monitoring (CM SAF) [5] the optimal angle for setting photovoltaic panels for this latitude will be a slope angle of $\alpha = 35°$. The monthly in-plane irradiation for a fixed angle is shown in Fig. 4.

The power of one photovoltaic panel is determined by the following:

$$P_{PV} = \eta \cdot A \cdot I_\alpha \tag{3}$$

where:

η – PVs panel efficiency,
A – collector active Surface area,
I_α – global irradiation at optimal slope angle.

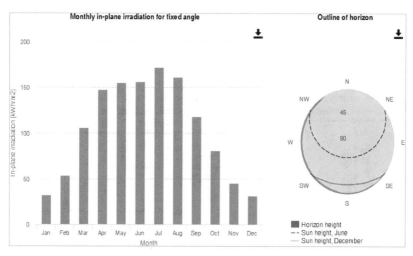

Fig. 4. Monthly in-plane irradiation for fixed angle [6].

3 Configuration of Wind – Solar Hybrid Power Generation System

To optimize the hybrid solar – wind power plant, the General Morphological Analysis (GMA) [7–10]. The entire process of carrying out the GMA is presented in [11].

Based on GMA, a hybrid solar-wind system consisting of PV panels. A 500 W wind turbine was selected, a three-propeller with a rotor diameter of 3 m. The limit and take-off wind speeds are 35 and 2 m/s, respectively. It is assumed that the turbine will be equipped with a mechanical locking system and a three-phase asynchronous motor. The efficiency of the generator is 96%. 80 photovoltaic panels about 600 W power and efficiency of 19.7% will be used to power the household each. The photovoltaic panels will be positioned at a 35° slope angle. The interconnected panels are then connected to the battery charging controller. It should be noted here that the technical data of the photovoltaic panels given in the catalogue cards are measured under STC conditions, which means that in order to obtain, during the analysis, a constant air temperature of 25 °C would have to be maintained throughout the year. An increase in the number of photovoltaic panels could be a solution to increase the energy produced. The entire system will be equipped with 4 batteries with a capacity of 150 Ah, an overload band power of 660 W and a rated voltage of 12 V. Each of the modules – wind and solar – will be connected to the controller before connecting to the battery, the task of which will be m.in. Protection of batteries from overcharging.

The block diagram of the designed hybrid system is presented in Fig. 5. On its basis, a numerical model of the hybrid power plant was developed in the MATLAB®/Simulink software, as shown in Fig. 6.

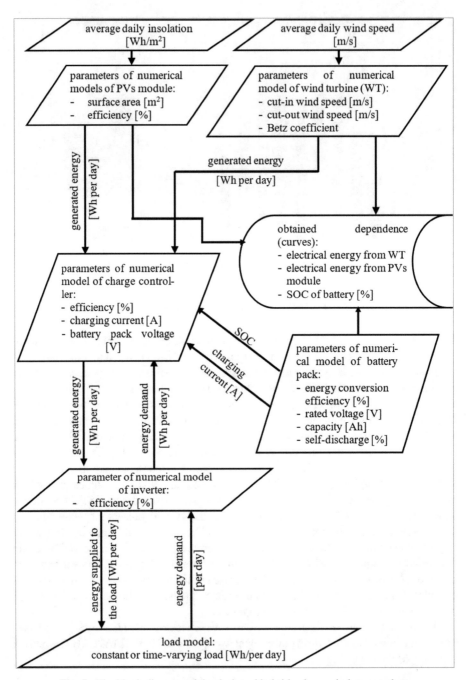

Fig. 5. The block diagram of the designed hybrid solar – wind power plant.

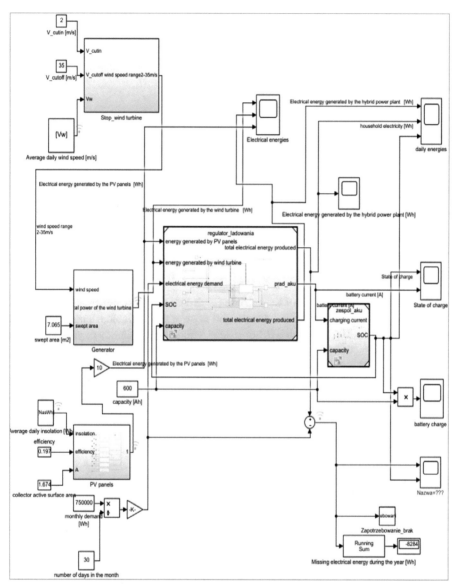

Fig. 6. The block diagram of the numerical model of a hybrid solar -wind power plant developed in MATLAB®/Simulink.

4   Analysis of Results Obtained from Numerical Simulations

Graphs showing the amount of electricity generated by off-grid wind - solar hybrid power plant and its individual components, i.e. the wind turbine and the eighty PV panels on individual days during the year, are shown in the figures Fig. 7, 8 and 9 respectively.

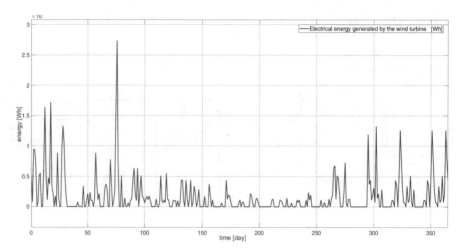

Fig. 7. The graph of the amount of electrical energy generated by the wind turbine.

When analysing the graph of the amount of electrical energy generated by the wind turbine, it can be seen that this characteristic is strongly variable, mainly dependent on wind speed (see Fig. 3), which is consistent with the dependency expressing power contained in the air stream (Eq. 1).

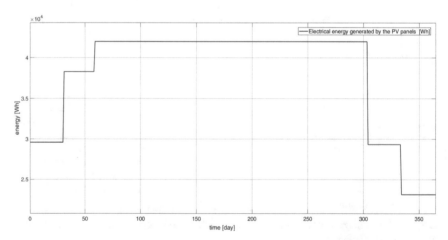

Fig. 8. The graph of the amount of electrical energy generated by the PV panels.

Analysing the graph for the PV panels, which is shown in Fig. 8, it can be noticed that the forecast value of electricity generated for most of the year, i.e. from March to October, is at 4.2 kWh of energy produced on a one-day scale. This value is due to the maximum power parameter of the photovoltaic panel, the value of which was assumed in the simulations as 600 W. Unfortunately, this is one of the higher power panels offered by manufacturers. Additionally, these panels are characterized by a quite high efficiency compared to others, equals to 19.57%.

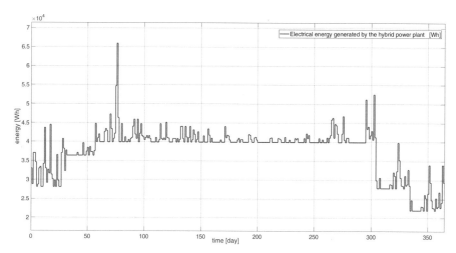

Fig. 9. The graph of the amount of electrical energy generated by the hybrid power plant.

However, the graph of the amount of total electrical energy generated by the hybrid power plant is shown in Fig. 9. It is a state in which the home power grid would use all the electricity produced. Then there would be no need to store surplus electricity in batteries because the amount of energy consumed by the home network would not be limited by the value of demand.

The most important characteristic obtained from the conducted numerical simulations is the characteristic representing the amount of energy that is allocated to power the household. It is the result of the operation of a wind turbine, photovoltaic panels, batteries, and the charging regulator controlling the entire system. This graph is shown in Fig. 10. In December, the household's electricity demand is 817 kW. From the designed the hybrid power plant, 730.76 kW of electricity is obtained, and therefore 10.55% of the demand for it is not covered.

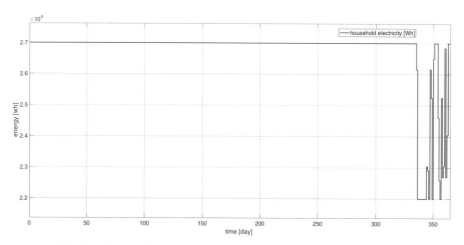

Fig. 10. The graph of the amount of electricity supplied to the household.

The comparison of the total amount of energy produced by the hybrid power plant with the shortage of electricity supplying the household is presented in Fig. 11.

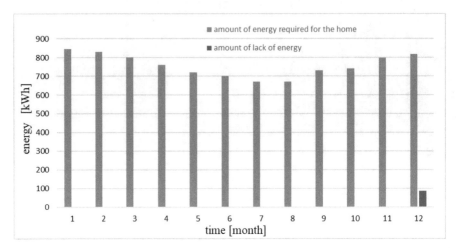

Fig. 11. The comparison of the total amount of energy produced by the hybrid power plant with the shortage of electricity supplying the household.

The analysed the hybrid power plant was equipped with batteries, the task of which is to store surplus from the electricity produced in order to supply the household on days during the year, when the amount of total electricity produced by the hybrid power plant is too small. As it is an expensive solution, special attention should be paid to the operation of batteries. Two characteristics are presented: the change in the State of Charge battery (SOC) level is shown in Fig. 12; the battery current is shown in Fig. 13.

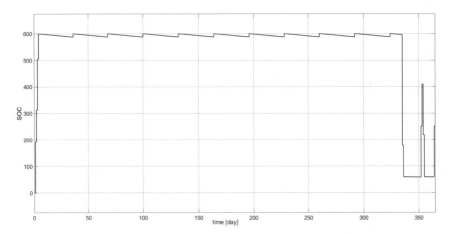

Fig. 12. Battery operation characteristics: State of Charge parameter (SOC).

Based on the graphs in the Fig. 12, you can see how often the surplus electricity is stored in the batteries, and how long the batteries remain charged.

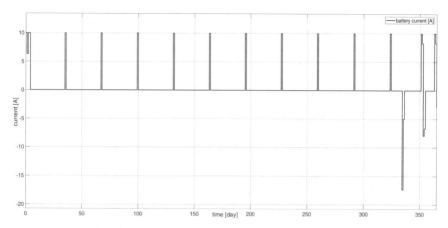

Fig. 13. Battery operation characteristics: battery current.

The diagram shown in Fig. 13 shows the course of the battery current value. Each falling slope means that the energy stored in the batteries was used to power the farm, while the rising slope means that on a given day the amount of energy was enough to power the entire farm and energy stored in the batteries.

5   Conclusions

Based on the data resulting from numerical simulations conducted for the designed hybrid solar-wind power plant, supplying a household, several conclusions can be drawn.

The amount of electricity generated by the solar part, i.e. PV panels, for solutions currently available on the market, is constant. This means that the current technical solutions are not able to fully use the energy potential of the sun.

On the other hand, the energy produced by the wind part of the power plant is very variable throughout the year. Its value depends primarily on the wind speed. Despite the low average values of wind speed in the area in question, the obtained results are satisfactory.

Based on the data resulting from the simulations for the adopted model of a solar-wind hybrid power plant supplying a household, it can be concluded that the amount of missing energy, i.e. the difference between the demand and the energy generated by the modelled system, amounting to 10.55%, is a very satisfactory result. Therefore, observing the further development of both photovoltaic cells and batteries, such energy sources for countries with not very favourable climatic conditions, for example for Poland, should be treated as a viable alternative to non-renewable energy sources.

Energy storage is a very expensive solution, and it did not bring satisfactory results for the analysed configuration. The batteries, for the input data used, only reached the state of charge a few times, and the stored energy was used very quickly. Such use causes the batteries to wear out very quickly. Energy surpluses could be safely stored in a different way, e.g. in water tanks, which would be a cheaper and more ecological solution. The resignation from the storage of surpluses in batteries would not bring large losses in this case.

When considering the installation of such complex solutions, the parameters of all system components should be selected in such a way that they meet the assumed results to reduce losses.

References

1. Calculator for electricity consumption at home. https://kalkulator.tauron.pl/h5/
2. Public data of the meteorological station in Katowice. https://meteo.gig.eu/
3. Benghanem, M.: Optimization of tilt angle for solar panel: case study for Madinah. Saudi Arabia. Appl. Energy **88**(4), 1427–1433 (2011)
4. Landau, Ch.R.: Optimum tilt of solar panels. https://solarpaneltilt.com/optsolar.html
5. Satellite Application Facility on Climate Monitoring (CM SAF). www.cmsaf.eu
6. Photovoltaic geographical information system. https://re.jrc.ec.euroa.eu/pvg_tools/en/#PVP
7. Ritchey, T.: General morphological analysis as a basic scientific modelling method. Technol. Forecast. Soc. Change **126**, 81–91 (2018). https://doi.org/10.1016/j.techfore.2017.05.027
8. Ritchey, T.: Problem structuring using computer-aided morphological analysis. J. Oper. Res. Soc. Special Issue Problem Structuring Methods **57**, 792–801 (2006)
9. Ritchey, T.: General morphological analysis, a general method for non-quantified modelling. Paper presented at the 16th EURO Conference on Operational Analysis, Brussels (1998)
10. Ritchey, T.: Wicked Problems –Social Messes. Decision Support Modelling with Morphological Analysis, 105 p. Springer, Heidelberg (2011)
11. Jureczko, M.: Optimization of stand-alone hybrid solar - wind system by using general morphological analysis. In: Maalawi, K. (ed.) Modeling, Simulation and Optimization of Wind Farms and Hybrid Systems. IntechOpen (2020). ISBN 978-1-78985-612-5, Print ISBN 978-1-78985-611-8 eBook (PDF) ISBN 978-1-78985-645-3. https://doi.org/10.5772/intechopen.90284

The Automatic License Plate Recognition System Based on Video Sequences Using Artificial Neural Networks

Mariola Jureczko$^{(\boxtimes)}$ and Filip Uherek

Department of Theoretical and Applied Mechanics, Faculty of Mechanical Engineering, Silesian University of Technology, 18A Konarskiego Street, 44-100 Gliwice, Poland
Mariola.Jureczko@polsl.pl

Abstract. The current development of technology provides us with access to increasingly complex computers with enormous computing power. Reducing their dimensions and increasing availability on the market resulted in a reduction in their manufacturing costs. Occurred situation causes increased development of applications based on artificial intelligence methods. On the other hand, the user uses a web browser to access the computer with computing power much greater than his desktop computer. Such possibilities are offered to us by the most popular platform offering free computing power and the ability to store data which is Google Collaboratory. Greater computing power significantly speeds up the complex calculations required to train artificial neural networks, which are becoming more and more popular.

The purpose of the project presented in this article was to develop an application for identifying - reading based on video sequences - car license plates. One of the machine learning methods and the deep teaching method using convolution neural networks were used to achieve the set aim.

Keywords: Automatic license plate recognition · Artificial neural networks · Automatic number – plate recognition · Convolutional Neural Networks

1 Introduction

Technological progress introduces new trends and solutions in science. In recent years, Artificial Neural Networks (ANNs) used for deep learning have increased its popularity. They are being used in various fields, from medicine to business, through the production and improvement of machine construction.

ANNs allow you to give computers human features. Currently, the device can understand our speech, writing, and even detect our movements. What is more, the way these abilities are given to the computer does not force the hiring of crowds of programmers creating extensive programs containing thousands of lines of code.

The way ANNs work is inspired by the action of neurons in the human brain. Inspiration based on nature has been guiding science for years. Drawing behaviour patterns from the environment and transferring it to the digital world, one can conclude that with appropriate external stimuli repeated many times, even very complex tasks will become simple and natural for man. The computer program based on neural networks works in a very similar way. Properly prepared data is fed to the input neurons, where the learning process takes place. Depending on the level of difficulty and the complexity of the problem, the appropriate number of layers and neurons are used. Information processed through layers enables neurons to adapt to data and find patterns occurring in given set. After enough transitions, the output neurons can return feedback based on observations acquired during learning process [1–4].

The above-mentioned features have one very significant disadvantage. Since neural networks receive only the information contained in the data, providing incorrect data might result in poor outcomes. For example, in image analysis based on neural networks, models trained to classify images of cats and dogs will be able to recognize only these two objects. While trying to categorize a photo of e.g. car's, the application will inform that the detected object is an animal. It happens because the information provided as the input to the network causes the activation of the neurons, which are responsible for determining the belonging of data to the class responsible for cat-type objects, or dog-type objects. This trivial example illustrates how powerful and dangerous the ANN's are.

2 Overview of the ALPR System

Automatic plate recognition systems exist under many names [5]: Automatic Number Plate Recognition (ANPR), Automatic License Plate Recognition (ALPR/LPR), and Car Plate Recognition (CPR). In each case, we are dealing with systems of the same type.

One of the most popular and oldest methods of "artificial intelligence" based on image analysis is the ALPR system. The first system under the given name was created in 1976 in Great Britain. The system was developed by the police research and scientific centre. At first, it was implemented in only two locations, and its first success was the detection of stolen vehicles [6]. Currently, the mentioned system is used on every highway, vehicle control point, borders, speed cameras, and even in shopping centres. It is also used for statistical analyses, aimed at, among others to determine the average speed on the designated section of the route. In contrast to speed cameras, which are used to record speeding in each place. The ALPR system implemented in a controlled point of the section of the speed measurement system allows us to verify the time in which a given vehicle covered a given section of the road. The system calculates the average speed and compares the obtained result with time-limit thresholds permitted for the given section of the route [7]. In addition to the wide application allowing for quick and accurate detection of law violations, the ALPR system is also used to improve parking lots in shopping centres. The plate recognition system significantly increases their throughput, eliminating the need of using parking tickets. The parking fee is regulated in free-standing machines by entering the characters contained in the car's registration plate, which are stored in databases.

A typical the ALPR system consists of [6]:

- high - quality camera,
- Optical Character Recognition (OCR) software for character recognition,

- computer – the central unit responsible for analysing and recognizing characters contained in the license plate,
- built-in internal memory,
- image gathering database,
- transmission module.

Despite the many benefits of using the ALPR system, it does not guarantee 100% reliability. Under real conditions, a modern and well-functioning system can to correctly identify 90–94% of all license plates registered vehicles [6]. Weather conditions, contamination, or deformation of license plates, unusual, have a significant impact on its correct operation location of the board, insufficient lighting. All the above-mentioned defects hinder the video analysis system from functioning properly and may result in incorrect verification of recognized characters, or extreme cases, complete failure to verify.

3   Development of the ALPR System Application

The block diagram containing the individual stages of the application is shown in Fig. 1 and Fig. 2. The purpose of the application is to detect and recognize the characters contained on the vehicle's license plate. Using blocks, the logic of the developed application was presented. The application operation diagram is reduced to analysing individual frames obtained from the video sequence.

The application architecture is as follows. The *start* block is a block that initializes the application. The way of application initialization may vary depending on the interpreter used. The video sequence which potentially contains vehicle frames is transferred to the *video_frame* operating block. The *video_frame()* function uses the *VideoCapture* class available in the OpenCV library, which allows for splitting video sequences into frames. Successive frames are saved to a list which is presented asset of all frames making up the video sequence. The *frame* input/output block, shown in Fig. 1, refers to a single frame obtained from a video sequence.

Then subsequent operations in the discussed application subject the extracted single frame from the video sequence.

Frame analysis via the *cnn_frames()* function causes high CPU usage. It is due to the artificial neural network model used at this stage of the application. The use of a given model involves the detection and localization of vehicles. To isolate the area of the frame containing the vehicle. Previously trained Tiny YOLO (You Only Look Once) model was used at this stage of the application [8–10]. The modern real-time object detection system, Tiny YOLO, is based on one neural network made of a convolutional neural network. The neural network divides the image into smaller areas and then predicts the areas and the probability of occurrence of the object for each of the designated regions. The designated areas are then analysed in terms of the probability of occurrence of the searched object in each region [8–10].

98   M. Jureczko and F. Uherek

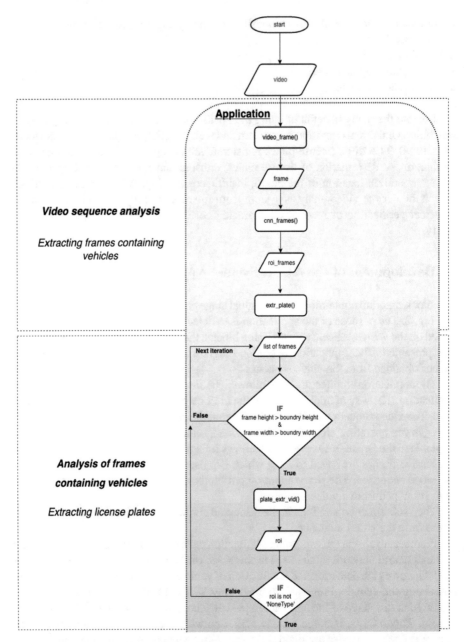

Fig. 1. Block diagram 1st part.

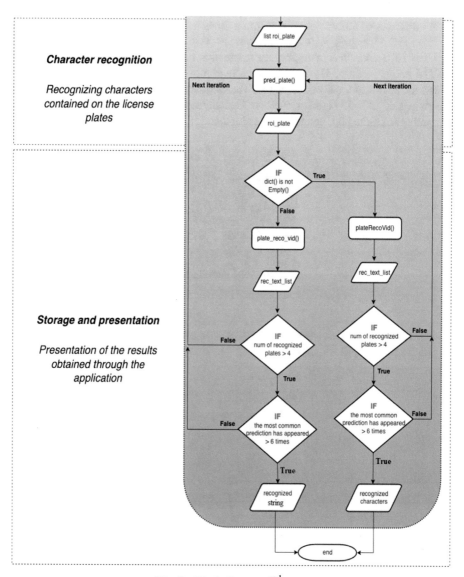

Fig. 2. Block diagram 2nd part.

An example of a frame obtained from a video sequence is shown in Fig. 3.

The YOLO neural network receives as the input a tensor with dimensions of (416, 416, 3), which is in the form of frame obtained from a video sequence. Frame is an RGB image made of 416×416 pixels. In contrast, the output of the neural network returns the tensor of the dimensions (13, 13, 125). Tensor has the form of a grid with the dimensions of 13×13 cells imposed on the analysed image. Detailed description of the interpretation of the obtained tensor was included in [11].

Fig. 3. Frame obtained from the video sequence.

A grid with dimensions of 13 by 13 cells is applied to the frame obtained from the video sequence, as shown in Fig. 4.

Fig. 4. Frame with an imposed grid.

The image classification and location process are applied to each cell. The Tiny YOLO neural network predicts the probability of the occurrence of the searched object in each cell and determines the coordinates of the frame containing the expected object.

Assuming one gird cell might contain more than just one detected object Tiny YOLO returns 5 predictions for each grid cell. Given this information, output tensor can be considered as a tensor of dimensions of (13, 13, 5, 25) where 13 by 13 refers to the dimension of the imposed mesh, 5 designates to the number of predictions per grid cell and 25 stands for the vector consists of 20 values responsible for the probability of recognizing a given classes (Tiny YOLO has been trained to recognize objects of 20 classes) and 5 values responsible for the probability of occurrence of the object and the coordinates of the frame containing the object. Thus, the neural network detects ($13 \times 13 \times 5 = 845$) 845 frames.

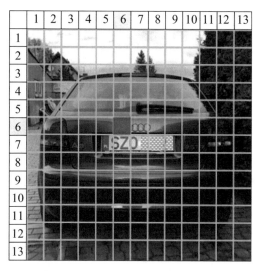

Fig. 5. Frame with a selected grid cell (red colour).

Considering one grid cell (for example grid cell (6,6) as shown in Fig. 5) the output of the neural network model for a given grid cell is a vector y with a length of 25.

$$y = (p_c, b_x, b_y, b_h, b_w, c_1, \ldots, c_{20}) \tag{1}$$

where: p_c stands for probability of detection object in the given window, while the following, b_h, b_w values are responsible for frame coordinates as is shown in Fig. 6 and b_x, b_y values are responsible for grid cell coordinates as is shown in Fig. 5 and Fig. 6, and c_n values correspond to the class of an object.

Fig. 6. Obtained bounding box.

The source code used in the *cnn_frames()* function responsible for retrieving and decoding information received as the output from the neural network was taken from [12]. To limit the number of classes recognized by the Tiny YOLO neural network, changes have been made to the source code. A restriction was introduced to the function to filter the results obtained to select only objects of class: bus, car, and motorbike. Modification of the source code also involved changing the value returned by the function by adding the *roi_frames* variable to the return statement (see Fig. 1). The newly created variable is in the form of a list with frames, which contain recognized objects cropped from the initial frame. The introduced changes guarantee protection in the case when there is more than one vehicle occurring in a single frame.

In the next stage of the application, the *roi_frames* list is passed to the *extr_plate()* function (see Fig. 1), in which the first verification of frames received using the neural network takes place. Given operation is presented in Fig. 1 in the form of a decision block that filters the following frames by checking their size. In the application, 200 pixels were used as the limit of the frame height and 340 pixels as the limit of frame width. All frames with dimensions smaller than those specified by the boundary values are ignored. Frames that meet the criteria of the decision block condition are passed to the *plate_extr_vid()* function. The task of the *plate_extr_vid()* function is to determine and crop from the frame the area containing the license plate. For this purpose, the frame process is performed using the image processing functions available in the OpenCV library, consisting of the following steps:

1. The use of Gaussian blur operations to reduce image noise [13]. The effect of Gaussian blur operations is shown in Fig. 7.

Fig. 7. Image after Gaussian blur operations.

2. Thresholding operation to obtain a binary image. The effect obtained after performing image thresholding operations is shown in Fig. 8.
3. Operation to determine the contour. The *cv2.findContours()* operation was used for this purpose. The operation of determining the contour consists of connecting the curve of all points (pixels) having the same colour or intensity (Fig. 9) [13]. The resulting contours are presented in the form of rectangles, which are finally imposed on the frame (Fig. 10).
4. The final stage of the *plate_extr_vid()* function, consists of verifying the rectangles (contours) obtained, by checking their dimensions. As a result of the operations described above, the license plate is extracted from the frame as shown in Fig. 11.

Fig. 8. Image after thresholding operation.

Fig. 9. Determined contours.   **Fig. 10.** Contours in the form of rectangles.

Fig. 11. Extracted license plate.

The application's next step is to transfer the image representing the license plate to the *pred_plate()* function (see Fig. 2). One of the difficulties encountered when implementing the automatic system for recognizing license plates from video sequences was solving the problem of multiple appearances of the same vehicle in the analysed video sequence. This problem particularly concerns the implementation of a mobile type of the ALPR. A vehicle equipped with this type of system is exposed to a situation in which, while the system is active, the video generated by the camera may contain a frame representing the same vehicle for a longer period. For this purpose, a dictionary structure was implemented in the *pred_plate()* function, which allows creating keys and assigning different types of data to them. The first step in the *pred_plate()* function is to verify the dictionary status. This process using a decision block controlling the content of the dictionary is shown in Fig. 2. In case the dictionary is empty, i.e. during the process of application operation, when no predictions of the license plates have been made yet the decision block following the block called *roi_plate* (which contains extracted license plates) returns the value *False*, otherwise, when the dictionary is no longer empty, it returns the value *True*. Returning the value *False* leads to a part of the code where the frame representing the license plate is sent to the *plate_reco_vid()* function. This function implements the Tesseract OCR library, which has in its resources neural networks trained for character recognition (OCR) [14]. To increase the probability of correct character recognition, the image containing the license plate is exposed to multiple image thresholding operations and character recognition attempts using the *image_to_string* method called from the Tesseract OCR library.

The image of the license plate after the image thresholding operation, which granted the correct recognition of characters via the *image_to_string* method, is shown in Fig. 12.

Fig. 12. Characters recognition via *image_to_string* method.

Subsequent transitions consisting of changing the value of the *cv2.threshold* method performing the image thresholding operation generate different answers of the *image_to_string* method. To generalize the response of the final application, which is correct character recognition, all character recognition attempts are saved to the *rec_text* list. Then the list is verified in terms of its content. If the list contains a minimum of four successive attempts to recognize characters and the most common prediction appears in the list more than four times, a new key is created in the *predPlate* dictionary. The key in the dictionary refers to the list containing the license plate image and recognized characters.

Subsequent iterations in the application process boil down to comparing the saved information in the dictionary with the newly obtained license plates. The decision block following the block named *roi_plate* then returns the value True.

Value *False* is returned when the dictionary is empty, which causes part of the code to be called, in which the frame representing the license plate goes to the *plateRecoVid* function. This function uses the Tesseract OCR library [14, 15], which has in its resources neural networks trained to recognize characters. To increase the likelihood of correct character recognition, the image containing the license plate is subjected to multiple image thresholding operations and character recognition attempts using the *image_to_string* method called from the Tesseract OCR library.

Returning the value *True* calls, the part of the code in which the newly obtained license plate is compared to the one previously saved in the *pred_plate* dictionary. For this purpose, the *feature_comp()* function was used. The exact operation of this function was described in the article [16]. The feedback of the *feature_comp()* function is a *True* value if the newly analysed license plate image is identical to the image from the previous iteration from which the characters were recognized.

A positive response of the *feature_comp()* function calls the *plate_reco_vid()* function again and the character recognition process occurs in the same way as described above in the previous paragraphs. An element protecting against excessive use of the system resulting from repeated recognition of an identical license plate was implemented by the control of the content behind the current *pred_plate* dictionary key. If more than 3 successive character prediction attempts have been assigned to a given key, a new key is created and the character recognition operation for the analysed license plate image is considered completed.

Each consecutive iteration containing the same license plate is skipped. However, the appearance of a new license plate triggers an identical process to the one described above.

4   Conclusions

The project discussed in this article aimed to develop a proprietary application for identification - reading based on video sequences - license plates of cars in motion.

At the design stage, as a part of the developed algorithm, the source generating data in the video sequence format was the Raspberry Pi/Arduino Uno microcontroller expanded with a camera module that sent video sequences to a unit with more computing power for further analysis. In the next steps/stages there is a specific extraction of video recording leading to the extraction, separation of relevant information. This application works on the principle of protocol processing, using available libraries, compatible with the Python language.

The first stage of the algorithm is to detect and extract the desired objects from the video sequence. For this purpose, previously over-trained Convolutional Neural Networks (Kerras/TensorFlow libraries) were used. A positively classified object is saved in a single frame format and forwarded to the next stage of the program.

The second stage involves applying image filters using OpenCV libraries to extract the area, region of the image containing the license plate.

After correctly determining the region of interest, the next step is to recognize individual characters. Recognition is carried out using the proprietary OCR-based model. Finally, the visualization and final presentation of the acquisition result - the vehicle's registration number.

Image analysis based on artificial neural networks features new possibilities that would be unattainable using traditional image processing operations. The only obstacle resulting from the use of artificial neural networks is the need for large data sets with correct labelling.

Verification of performance of the application was carried out using a laptop with an *Intel i5-4300M CPU 2.60 GHz* processor. As test data, a video sequence of 139 frames displayed at 25 frames per second was used. The results of the tests carried out were correctly recognized license plates. The application speed was 0.62 s/frame.

The main problem limiting the functionality of the application is weather conditions. The presence of heavy rainfall, snowfall, or limited visibility due to fog or lack of daylight significantly reduces the effectiveness of image analysis. Much better performance can be achieved under constant conditions, e.g. in underground garages, where the lighting remains constant.

References

1. Mohamad, H.H.: Fundamentals of Artificial Neural Networks. MIT Press, Cambridge (1995)
2. Tariq, R.: Make your own neural network. Kindle Edition (2016)

3. Hagan, T.M., Demuth, B.H., Beale, H.M., Orlando De J.: Neural Network Design, 2nd edn. Martin Hagan (2014)
4. Korbicz, J., Obuchowicz, A., Uciński, D.: Artificial Neural Networks: Basics and Applications. Academic Publishing House PLJ, Warsaw (1994). (in polish)
5. Stroński, M.: ALPR (Automatic License Plate Recognition) systems capabilities and use in traffic control systems. Bulletin KLIR No 72, 61-66 (2011). (in polish)
6. Matysiak, A., Kruszewski, M., Niezgoda, M., Kamiński, T.: The analysis of ANPR camera location points in bus lanes monitoring system in the city of Warsaw. J. KONES Powertrain Transp. **20**(3), 269–275 (2013)
7. Jonaszek, D.: Section measurement of speed. Metrology and assaying. Bull. Central Office Meas. **4**(7) (2014)
8. Redmon, J., Divvala, S., Girshick, R., Farhadi, A.: You only look once: unified, real-time object detection. In: Proceedings of the IEEE Conference on Computer Vision and Pattern Recognition (CVPR), pp. 779–788 (2016)
9. Menegaz, M.: Understanding YOLO (2018). https://medium.com/hackernoon/understanding-yolo-f5a74bbc7967
10. Redmon, J., Farhadi, A.: YOLO9000: Better, Faster, Stronger. arXiv preprint arXiv:1612.08242 (2016)
11. Hollemans, M.: Real-time object detection with YOLO. https://machinethink.net/blog/object-detection-with-yolo/
12. Source code of the postprocess function. https://github.com/thtrieu/darkflow/blob/master/darkflow/utils/box.py. https://github.com/allanzelener/YAD2K/issues/3. Accessed 9 Dec 2019
13. OpenCV library documentation. https://docs.opencv.org/4.1.2/d1/dfb/intro.html. Accessed 9 Dec 2019
14. Tesseract OCR. https://github.com/tesseract-ocr/tesseract. Accessed 9 Dec 2019
15. Kay, A.: Tesseract: an open – source optical character recognition engine. Linux J. **2007**(159) (2007). Belltown Media, Houston
16. Feature Comp function. https://opencv-pythontutroals.readthedocs.io/en/latest/py_tutorials/py_feature2d/py_matcher/py_matcher.html. Accessed 9 Dec 2019

Analysis of the Form the Constructional Orthosis Sloppy Foot

Paweł Jureczko$^{(\boxtimes)}$ ⓘ and Magdalena Kogut

Department of Theoretical and Applied Mechanics, Silesian University of Technology,
ul. Konarskiego 18a, 44-100 Gliwice, Poland
`Pawel.Jureczko@polsl.pl`

Abstract. The paper describes the construction of sensory inserts on footwear with measuring elements that measure the pressure exerted by the foot on the ground. The process of creating a prototype in 3D technology was also presented and measurements were made to verify the correctness of the presented solution.

Keywords: Orthosis · Sloppy foot

1 Introduction

Stabilization devices play a key role in the process of rehabilitation of musculoskeletal system. They can immobilize and stabilize the limbs and they also help relieve the pressure on them. They also ensure correct positioning of the limbs, thus contributing to pain reduction or prevention. One such device is an orthopedic denture, which is intended to secure a specific part of the body after an injury or surgery and not to lead to further damage. A great advantage of the braces and their decisive advantage over plaster is the fact that they are lightweight and more comfortable than plaster, thanks to which wearing them is not cumbersome and allows relative mobility. However, we must remember that immobilization has a big impact on muscles, bones and ligaments and tendons. Atrophy of the muscles occurs, which leads to a period of restoring mobility twice as long as the stabilization time. This is a result of lack of muscle activity and normal strain. The structural properties of ligaments and tendons also weaken and collagen metabolism increases. The density of the bone decreases, as does its stiffness. However, it all depends on the place of stabilisation, the length of the stabilisation and the age of the animal, as bone atrophy progresses faster in dogs during the growth period. Bone reconstruction is a very difficult process, as it is ten times slower than bone atrophy That is why it is important to know the effects of immobilisation when using stabilisation devices [1–4].

2 Analysis of Available Solutions

The solutions currently used in stationary and online shops are not patented by their manufacturers. This is because detailed information about patented products is available on patent offices'; websites or in search engines such as Google Patents. Manufacturers do not want to bear the cost of product claim. For this reason, the following patent analysis includes solutions previously used to stabilize animal limbs, which show how the design of stabilizers has changed over the years. The first of these is a joint stabilization device consisting of elements stretching between the bands attached to a limb above and below the joint. It can be additionally equipped with a structure limiting the range of movement in order to better adapt the position of the limb to the stage of rehabilitation. This stabilizer is lightweight and functional, it also allows you to change dressings after injuries. The solution described is shown in Fig. 1.

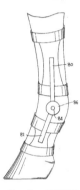

Fig. 1. Patent application US20140148746A1 [5].

Another solution is a stabilizer made of two layers. The first is the binding layer, which aims to stabilize, relieve strain, immobilize and guide the limb accordingly. The protective layer prevents dirt, water and moisture from penetrating the main unit. The Velcro fastener is fastened with a locking device to protect the animal's limb as much as possible. This is shown in Fig. 2.

Fig. 2. Patent application DE102012111102A1 [6].

An interesting solution is the dog's foot stabilizer shown in Fig. 3. It is considered in the analysis carried out due to the rotation of the animal's foot. It is made up of two parts connected together by a stabilizing strap, which can be separated if necessary. The first one wraps around the paw and the second one wraps around the bottom of the limb.

Everything is made of a flexible material with a high coefficient of friction on the outer surface. The agent contains a layer of foam.

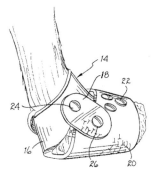

Fig. 3. Patent application US6546704B1 [7].

Another solution is a restraining rail that can be easily adjusted to the length of the animal's limb. It has through holes to allow ventilation of the railed area and is transparent, which allows you to control the condition of the bandage and limb without removing it. The solution described is shown in Fig. 4.

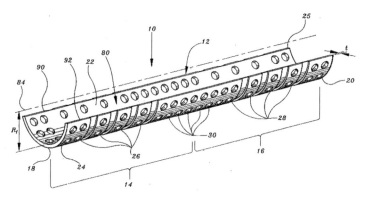

Fig. 4. Patent application US5134992A [8].

Another patented solution is a rail made of the main section made of plastic, in the shape of a semi-cylinder. The inner surface is lined with soft and resilient material. The splint allows the wrist joint to be placed in the resting position and also contains appropriate cavities around the wrist pad to prevent necrosis. This can easily be seen in Fig. 5.

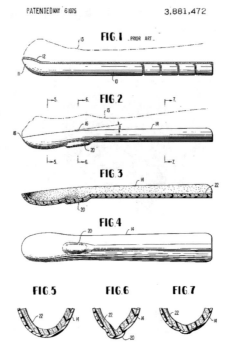

Fig. 5. Patent application US3881472A [9].

The veterinary device shown in Fig. 6 is used to adjust the limb bend. It allows you to achieve the right alignment with two sections. The first is mounted on the elbow section and the second on the shoulder section. Both are connected so as to transfer pressure to the elbow joint, thus enabling the stabilization of the limb.

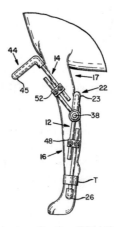

Fig. 6. Patent application US4440159A [10].

Another patent application concerns a rail set with intravenous fluid supply tube. The rails allow you to adjust the device to the length of the animal's limb, and an additional latch allows you to attach the set to the door of the transport cage. This is shown in Fig. 7.

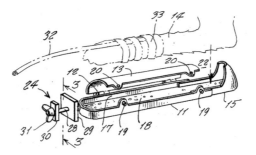

Fig. 7. Patent application US4505270A [11].

The presented solutions show different ways to stabilize the limb. Some of them consist of the rail itself, others have additional elements to limit the bending angle, for example. As the patent analysis shows, attention should be paid to the proper contouring of the device body and its inner layer. This increases the comfort of use and speeds up the rehabilitation process and prevents other injuries. Continuous development on the limb stabilizer market is visible not only in the case of foreign companies. Also in Poland, innovative solutions are emerging, using new materials and ways of immobilizing the joints. An example is the use of a codura, polyester fabric which, thanks to the PVC coating, has waterproof properties. It is increasingly used in the manufacture of stabilisers due to its resistance to weather conditions and mechanical damage. The first one is a brace with a light and durable construction. It extends from the paw to the middle of the forearm to provide maximum support and maintain the correct position of the limb. It also allows easy access to wounds, and the light weight of the device increases the animals tolerance to the rail. The method of fastening the orthosis by using Velcro strips with additional inserts is shown in Fig. 8.

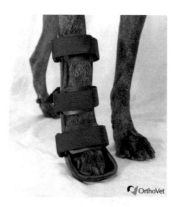

Fig. 8. OrthoVet splint [12].

An interesting solution for chronic joint injuries is the wrist bone protector shown in Fig. 9. Its strength is ensured by neoprene reinforced with plastic inserts. Velcro fasteners make it easy to fit the stabilizer to the leg. A very important feature of this product is the possibility to change its shape many times. Simply place the device in hot water for 30 s and when it has cooled down, place it on the animal's limb in its natural position. About 60 min later, the protector is ready for use. It is an ideal solution for long-term rehabilitation because it allows to adjust the shape of the inserts to the progress of convalescence.

Fig. 9. Karusse joint protector [13].

As the above review shows, an important aspect considered when making a stabilizing device is its weight. The device must not obstruct the animal's movements and must be too heavy. An important issue is also how to fix it. Velcro strips are used in all solutions. It is also good to use additional inserts that will not cause abrasions on the belts. It is also worth noting the materials of which the stabiliser is made. High quality materials guarantee a much longer and more comfortable use of the product. In order to make a stabilizer, manufacturers most often use synthetic gypsum in order to map the elements of the limb's structure and measure it. The form is obtained by wrapping a specific zone with a single layer of glass fibre dressing. You have to remember that the limb must be in a physiological position. Then the appropriate anatomical points should be marked on the casting surface, as well as the joint centreline. The next step is to cut the mould and pull out the limb, previously secured with a suitable film protecting the coat. Finally, the plaster bandage should be glued together.

3   Description of the Case

The correct positioning of the forelegs is shown in Fig. 10. By using stabilisation devices it is possible to achieve this position for animals with various injuries to the musculoskeletal system. The patient for whom the stabilizer design will be performed has many diseases of the right anterior paw such as: radial nerve paralysis, contractions of bending tendons, rotation of the paw inwards and partial sensation in the paw limited to two fingers. Due to his injuries, he is not able to position his limbs correctly while moving.

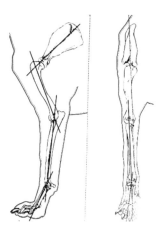

Fig. 10. Correct positioning of the forelegs [3].

In order to allow the patient to reach the appropriate paw position during movement, his limb should be stabilized from the humerus up to half of the radial bone. This will immobilize the elbow joint. In Fig. 11 between other things, the rotation of the paw inwards and the contraction of the flexor tendons in the patient can be observed. Unfortunately, the stabilizers available on the market only allow to take away the traffic possibilities in individual ponds. They do not provide a comprehensive treatment for every case, so it is better to make braces that take into account the individual needs of the patient. In this case, in addition to stabilizing the limb, it will be possible to combine the stabilizer with a DORSI-FLEX shoe to ensure the physiological position of the toes.

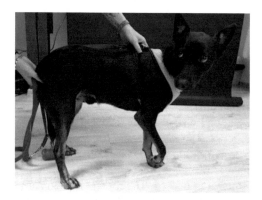

Fig. 11. The patient's anterior limb placement.

In the case of the design of a stabilisation device for the previously presented dog, one of the most important factors is the immobilisation of part of the humerus and elbow joint. The minimum length range of the stabiliser is determined by half the length of the radial bone to ensure the natural alignment of the animal's limb. Another important element is the wall thickness of the device. It must not be too wide, as it may cause your dog to be reluctant to use it. However, it should be noted that a wall that is too thin can be damaged as a result of daily use of the orthosis. The wall thickness should therefore be between 3 and 4,5 mm. It should also be remembered that this dimension will be increased by additional filling with an appropriate insert. The shape of the device also matters. A well-made geometry will better adapt to the animal's individual physique, thus increasing the functionality of the stabilizer. The device must also not be too heavy, obstruct walking, or have additional parts that will sag and endanger your dog. For this reason, attention should be paid to the method of attachment. The Velcro straps used shall have additional sleeves which will prevent the animal from removing the device from the limb. Also, the holes made for interlacing the tapes must not adversely affect the comfort of use of the orthosis. They must have a width that matches that of commercially available Velcro strips, but it is their location that plays a key role here. It is also important that the stabiliser is durable and does not deform excessively under the pressure of the animal's body. In the case of strength analyses, 100 N was taken as the loading force of the element. This is due to the dog's low weight (less than 20 kg) and the weight distribution during movement.

The dog is walking on the dog's feet, which are diagonally placed. This movement takes place alternately in pairs: left front foot and right rear, right front foot and left rear. Another important element is the model. The printing time of a solution also influences its evaluation. However, it is important to remember to compare concepts under the same conditions and using identical process parameters. In summary, the examples must be tailored to the active lifestyle of the animal, provide comfort, appropriate stability and must not cause injury.

4   Concepts for Dog Paw Stabilization Device

The following chapter presents three concepts of dog paw stabilization devices. They were used to select the solution best suited to the criteria, which were then refined further in the product design. Stabilizer models are made in Siemens NX 11.0 environment. Strength analyses were obtained using Inventor. It was assumed that the models would be made of acrylonitrile-butadiene-styrene copolymer (ABS). The load on the inner part of the stabilizing shell and adjacent edges with a force of 100 N and the application of a fixed support on the external surface of the model at the location of the humerus head was also taken into account. The supports are not located at the fixing points, as the results obtained in this way would mainly depend on the zones where the Velcro strips would run and the quantity of them. In addition, a grid with an average element size of 0,1 length of the stop frame is used. The minimum element is 0,2 of average size. The maximum angle of the grid triangle is 60°. The presented conditions were used to analyses all three concepts.

4.1   Concept I

The first concept shown in Fig. 12 is a stabilizer model that guarantees the immobilization of the limb at the required section. The wall thickness of the device is 4 mm. Four holes made in the brace reduce its weight and guarantee airflow. They also have another, very important function. The straps can be crossed through them with carabiners, which must be attached to the previously installed harness. In this way we get an additional way of fixing and increase the stability of the device. The notches in the area of the humerus head sufficiently affect the comfort of use. Velcro strap holders are also an important element. They extend beyond the area of the orthopaedic husk, allowing for thicker stripes, but can cause abrasions if the animal falls, lies down or stands up.

Fig. 12. The first concept.

Figure 13 and Fig. 14 show respectively visualizations of reduced stresses and displacements for the first concept.

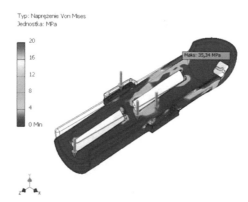

Fig. 13. The von Mises stress distribution [MPa] of the first concept.

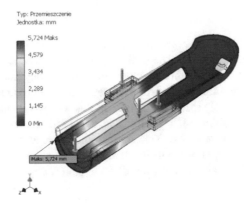

Fig. 14. The displacement distribution [mm] of the first concept.

As can be seen, the maximum reduced stress according to Huber-Mises-Hencky's hypothesis is 35,34 MPa and is located at the edge of the orthosis right next to the location of the fixed support. Higher stress values can be observed along the edges, almost to the locations where the handles are fixed and around the holes at the top.

4.2   Concept II

The second concept shown in Fig. 15 ensures that the limb is immobilised over a longer distance, thus guaranteeing greater stability when moving. It is properly profiled to achieve a physiological position. It has 4 mm thick walls and six holes for airflow and weight reduction. As in concept no. 1, additional fastening can be used here in the form of straps with carabiners interlaced through the holes. The holes for Velcro straps are made in the stabilizer shell. This prevents you from hooking up with them while they're moving. However, if thicker and less flexible straps are used, limb abrasions may occur, which will rub against the low-fitting straps during movement.

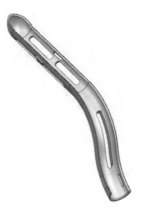

Fig. 15. The second concept

A visualisation of the reduced stresses is shown in Fig. 16. They are presented on a scale reduced to 20 MPa. The highest value, 51,13 MPa, occurs at the edge of the wall, right next to the fixed support. High values can also be seen in the holes for fixing Velcro strips and in the first pair of holes for interlacing strips with carbines. In the rest of the model, the reduced stress according to the Huber-Mises-Hencky hypothesis is up to 10 MPa.

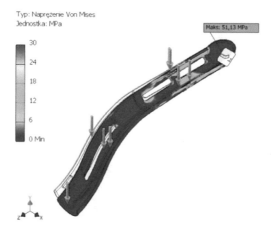

Fig. 16. The von Mises stress distribution [MPa] of the second concept.

You can see the displacement visualization made in Inventor in Fig. 17. The maximum displacement, 39,58 mm high, can be seen at the bottom edge of the device. Displacements above 30 mm can be observed throughout the farthest area of the stabiliser. High stresses and displacements result from the length of the device and the use of a fixed support on one extreme surface.

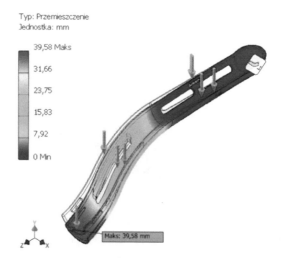

Fig. 17. The displacement distribution [mm] of the second concept.

4.3 Concept III

Figure 18 shows a third concept model made using Siemens NX 11.0. The solution includes a perfectly profiled edge around the humerus. Compared to other concepts, it also has thinner walls with a thickness of 3,4 mm, making it lightweight. The strap attachments protrude upwards so that the guided straps do not pose a hazard to the limb and there is also the possibility of using more durable straps. This position of the handles also allows them to be easily covered with a Velcro strap wrap, which limits the animal's access to them. To increase the fit of the stabilizer to the leg, you can also use straps with carabiners attached to the dog's harness on one side and to the handles on the other.

Fig. 18. The third design solution

In Fig. 19 and Fig. 20 one can observe successively visualizations of stresses according to the Huber-Mises-Hencky hypothesis and displacements. Stress is shown in the range from 0 to 15 MPa. The maximum value for a given case of 32,12 MPa is visible just below the rounded edge at the top of the stabiliser.

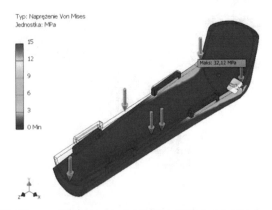

Fig. 19. The von Mises stress distribution [MPa] of the third concept.

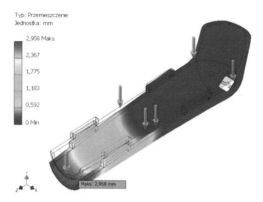

Fig. 20. The displacement distribution [mm] of the third concept.

Stresses in the range of 3 to 8 MPa can be seen next to the first pair of clamps. In the rest of the model the stresses do not exceed 2 MPa. The largest displacement, equal to 2,958 mm, applies to the lower edge of the device. At the top of the model the displacements shall not exceed 0,5 mm.

5 The Multi Criteria Analysis

The multi-criteria analysis is a way of selecting the solution that best meets the adopted criteria, which must be properly selected to ensure that the different options can be compared. It is also important that they have a significant impact on the functioning of the project. The following criteria, K1 to K7, were adopted for the following analysis:

- K1 - wall thickness of the model,
- K2 - print weight,
- K3 - print time,
- K4 - maximum value of reduced stress,
- K5 - maximum value of displacement,
- K6 - method of attachment (with particular attention paid to the inability to cause limb abrasion as a result of contact with Velcro straps and the occurrence of injuries associated with hooking the handles),
- K7 - functionality (expressed by the best fit to the shape of the limb in terms of stability).

The criteria shall be assigned the following weightings depending on the result of their comparison:

1 - for the more important criterion,
0.5 - for equivalent criteria,
0 - for the less important criterion.

The individual concepts, marked W1, W2 and W3, are assessed as follows:

0 - the concept does not meet the criterion,
1 - the concept meets the criterion sufficiently,
2 - the concept meets the criterion well,
3 - the concept fully meets the criterion.

A comparison of the options is shown in Figs. 21 and 22.

	K1	K2	K3	K4	K5	K6	K7	SUMA	W1	W2	W3	$W_{idealny}$
K1		0	0,5	0	0	0	0	0,5	2	2	3	3
K2	1		1	0,5	0,5	0	0	3	3	1	1	3
K3	0,5	0		0	0	0	0	0,5	3	1	2	3
K4	1	0,5	1		0,5	0,5	0,5	4	2	1	3	3
K5	1	0,5	1	0,5		0,5	0,5	4	2	1	3	3
K6	1	1	1	0,5	0,5		0,5	4,5	1	1	3	3
K7	1	1	1	0,5	0,5	0,5		4,5	2	2	3	3
Z1									41	26	56,5	63
d1									65%	41%	90%	100%

Fig. 21. Shows a graph indicating the extent to which the structural solutions developed are close to the ideal variant that fully meets all the criteria presented.

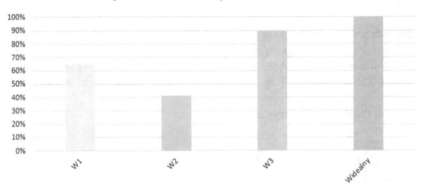

Fig. 22. Percentage assessment of individual solutions

As a result of the analysis, the most important criteria were determined, which are functionality and the way of fixing. Both rating a sum of 4,5 weights. Slightly fewer points were given to the maximum values of stress and displacement. The least important criterion in comparison with the others was the printing time. On the basis of a multi-criteria analysis, the solution that best meets the presented criteria was identified. This is the second concept. To determine where to make holes to reduce mass, a strength analysis of the stabilizer has been developed using Inventor. On the outer surfaces, on which the fixing tapes will run, fixed supports are placed. The inner part of the stabilizing hull and the adjacent edges are loaded, as in previous analyses, with a force of 100 N. The average size of the mesh element is 0,1 of the length of the stop frame and the minimum element is 0,2 of the average element size. The maximum angle of the triangular grid is 60°. The maximum stress is 5,457 MPa and can be observed at the lower pair of grips. In the rest of the model, the stresses are insignificant and do not exceed the tensile stress at the yield strength of 39 MPa for components made on a 3D printer.

As a result of the analysis, the most important criteria were determined, which are functionality and the way of fixing. Both collected a sum of 4,5 weights. Slightly fewer points were given to the maximum values of stress and displacement. The least important criterion in comparison with the others was the printing time. On the basis of a multi-criteria analysis, the solution that best meets the presented criteria was identified. This is the second concept. To determine where to make holes to reduce mass, a strength analysis of the stabilizer has been developed using Inventor. On the outer surfaces, on which the fixing tapes will run, fixed supports are placed. The inner part of the stabilizing hull and the adjacent edges are loaded, as in previous analyses, with a force of 100 N. The average size of the mesh element is 0,1 of the length of the stop frame and the minimum element is 0,2 of the average element size. The maximum angle of the triangular grid is 60°. The maximum stress is 5,457 MPa and can be observed at the lower pair of grips. In the rest of the model, the stresses are insignificant and do not exceed the tensile stress at the yield strength of 39 MPa for components made on a 3D printer.

For the boundary conditions shown, the reduced stresses and displacements are small. This means that in almost the entire component, apart from the immediate vicinity of the fixtures, holes can be made to reduce the weight of the device. The most commonly used solution is openwork construction. Unfortunately, due to the possibility of the animal's claws getting caught in small holes and the difficulty of removing them later in this case, this should be abandoned. Due to the good support of the limb, the material cannot be removed from the 30 mm wide belt running through the centre of the device. The holes must also not be too large, as this would cause the soft lining used to increase comfort. That is why two holes in the side walls were cut out in the model. Each of them is 60 mm long and 7 mm wide. Their edges are rounded with a radius of 2,9 mm. The holes allowed the weight of the model to be reduced from 52 g to 45 g, which was checked using Simplify3D. The maximum stress is slightly increased to 5,911 MPa at the edge at the second pair of handles. There are also a few places with stresses in the region of 0,6 MPa, especially in the case of rounded holes and clamp attachments. The highest displacement value changed by almost 0,03 mm and is equal to 0,04441 mm. It can now be seen in the band marking the center of the hole, right at its upper edge.

6 Summary

As a result of the analysis, the most important criteria were determined, which are functionality and the way of fixing. Both collected a sum of 4,5 weights. Slightly fewer points were given to the maximum values of stress and displacement. The least important criterion in comparison with the others was the printing time. On the basis of a multi-criteria analysis, the solution that best meets the presented criteria was identified. This is the second concept that will be discussed later in the paper. 3D printing can be used for rapid prototyping, which saves time and reduces the costs associated with processing the target material. Thanks to the use of 3D printing it was possible to make a prototype of the concept of the pressure sensor and to carry out a test verifying the correctness of operation. In case of need of redesigning, it is possible to change the CAD model and reprint it, which takes about 7 h. The use of the Finite Element Method (FEA) is very helpful in designing new solutions. It shows how a pressed measuring element will behave. This method allows to obtain colour maps of stresses, deformations and displacements. This allows the optimum cross-section to be selected by changing the geometric parameters of the beam in order not to exceed the maximum values for the material.

References

1. Coughlan, A., Miller, A.: Leczenie złamań u małych zwierząt. Wydawnictwo Galaktyka, Warszawa (2004)
2. Kassolik, K., Andrzejewski, W., Gilar, A.: Rozwój fizjoterapii weterynaryjnej w Polsce. Fizjoterapia, nr 17, s. 78–86 (2009)
3. Levine, D., Millis, D.L., Taylor, R.A.: Rehabilitacja psów. Elsevier Urban & Partner, Wrocław (2007)
4. del Pueyo Montesinos, G.: Fizjoterapia i rehabilitacja w weterynarii, przeł. Dawidowicz J., wydaw. Edra Urban & Partner, Wrocław (2017)
5. https://patents.google.com/patent/US20140148746A
6. https://patents.google.com/patent/DE102012111102A1
7. https://patents.google.com/patent/US6546704B1
8. https://patents.google.com/patent/US5134992A
9. https://patents.google.com/patent/US3881472A
10. https://patents.google.com/patent/US4440159A
11. https://patents.google.com/patent/US4505270A
12. https://www.orthovet.com/product/orthovet-standard-front-leg-splint/
13. https://www.firma-admiral.pl/3-szyny.html

Optimization of the Drilling Process in Multi-hole Objects

Jan Kosmol$^{(\boxtimes)}$ (iD)

Silesian University of Technology, 44-100 Gliwice, Poland
jan.kosmol@polsl.pl

Abstract. The paper discusses an engineering problem i.e. the machining of multiple holes in large-sized objects on a multi-spindle CNC machine tool (drilling machine). Regarding the very long time for development of such an object (a dozen to several dozens of hours) there is a need to optimize such technological operations. The optimization comes down to choosing the holes to be made in such a way as to make the most of all available spindles in the machine tool. This will mean the shortest possible time for all holes to be made. It is therefore a discreet optimization problem. The paper introduces the spindle use indicator (degree) SWWR and proposes a discrete function of a purpose, which represents the difference between the number of holes made by all LO spindles despite the last one, and the so called R Remainder, namely the number of holes made by the last spindle. For this function of the target we search for a natural number n, for which it takes a minimum value. This natural number allows to calculate the number of holes made by each spindle LO_{opt}, except for the last one, and this is the optimal solution. The paper presents numerical examples of such optimization.

Keywords: Optimization · Multi-spindle drilling machine · Multi-hole objects

1 Introduction

In mechanical engineering there are large-size objects with holes, often of the same diameter and length. It is a frequent situation that the number of these holes reaches hundreds or thousands of pieces. This is the case, for example, in the so-called heat exchanger partitions [1], whose diameters reach 3 m and more and the number of the same holes is a thousand and more. Figure 1 shows an example of such a partition.

The characteristics of such partitions are as follows:

- holes with identical diameters are distributed unevenly, i.e. symmetry axes cannot always be defined, although the graduation between adjacent holes is generally the same for all holes,
- on both sides, the holes have small chamfers, which facilitate the installation of pipes, which the heated medium (water) flows through,
- there are several such compartments in a typical heat exchanger and therefore the tolerances for their arrangement are narrow,
- the construction of such partitions is time-consuming, especially for holes, and therefore the most effective methods of their construction are sought.

© The Author(s), under exclusive license to Springer Nature Switzerland AG 2021
A. Mężyk et al. (Eds.): SMWM 2020, AISC 1336, pp. 123–133, 2021.
https://doi.org/10.1007/978-3-030-68455-6_11

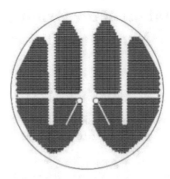

Fig. 1. Example of a heat exchanger partition

2 Subject Matter

Figure 2 shows a diagram of a typical multi-hole energy partition. Its characteristics are:

- the same diameter of all holes,
- equal graduation of all holes in the X-axis,
- equal graduation of individual hole rows in the Y-axis,
- a large number of holes, reaching up to a thousand and more.

The holes are usually made by drilling, on CNC numerically controlled multi-spindle machine tools (drilling machines) [2]. Such a machine has several independent head-stocks with WR spindles. Each headstock is individually numerically controlled, i.e. it is possible to program its movements along the X axis, in the range from L_{min} to L_{max}. This allows free positioning of the spindles in relation to the holes in the partition in one row. There is a minimum distance between two adjacent spindles that defines the minimum number of holes to be made by adjacent spindles.

The paper concern making holes in heat exchanger partitions by machining methods and on finding a drilling strategy. The strategy in this case is understood as the selection of the number of holes made by each spindle.

The main problem with this technology is to determine the coordinates of all holes in the XY system, as the drilling process will take place in this system of coordinates and selection of the order of drilling holes. It is therefore the issue of optimizing the technological process [3, 4].

Figure 2 shows an example of dimensioning a single hole row for an exemplary partition.

The principle adopted for dimensioning such objects and drilling holes on multi-spindle drilling machines is as follows:

- the dimensioning must be carried out separately for each hole row "i" ($i = 1$, $i = 2$, …)
- we dimension the coordinates in the X axis of each first hole made by individual spindles WR_1, WR_2, WR_3… WR_k, by entering the value $X_{i(1,1)}$, $X_{i(2,1)}$, $X_{i(3,1)}$, … $X_{i(k,1)}$,

- we dimension the coordinates in the X axis of each first hole distant from the previous hole by a distance L greater than the scale of the holes t (L > t, Fig. 2): $X_{i(1,2)}$, $X_{i(2,2)}$, $X_{i(3,2)}$, ... $X_{i(k,2)}$,

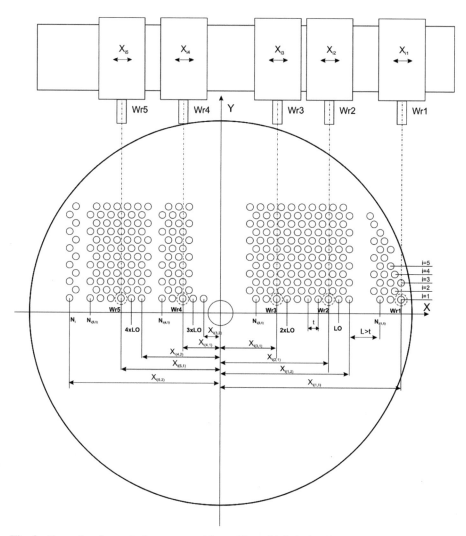

Fig. 2. Example of a typical energy partition with multiple holes: Wr_1, Wr_2, Wr_3, Wr_4, Wr_5 – symbols of spindles, N_i, $N_{i(1,1)}$, $N_{i(3,1)}$, $N_{i(4,1)}$, $N_{i(5,1)}$ – number of holes, LO – number of holes manufactures by one spindle

- we number the number of holes made by individual spindles WR_1, WR_2, WR_3, WR_3 ... WR_k, by giving the value $N_{i(1,1)}$, $N_{i(2,1)}$, $N_{i(3,1)}$, ... $N_{i(n,1)}$ counted from the first hole made by individual spindles up to the hole after which the distance to the next hole is greater than the graduation t (L > t),

- we assume that each spindle WR1, WR2, WR3... WR(n − 1), with the exception of the last WR_k, will make the same number of LO holes,
- each N_i hole row is described by a N_i matrix consisting of LWR records: WR1, WR2, WR3, WR4, WR5 .... WRLWR:

$$N_i = N_i(WR1, WR2, WR3, WR4, WR5, ... WRLWR) \qquad (1)$$

- each WR1 - WRLWR record is described by a system of parameters, namely (Fig. 3):

$$WRi = WRi(N_i, LO, t, N_{i(k,1)}, N_{i(k,2)}, N_{i(k,3)}, X_{i(k,1)}, X_{i(k,2)}) \qquad (2)$$

where:

N_i - the number of holes in the ith row,

LO - number of holes made by a single spindle,

t - standard graduation of the hole spacing in the X axis,

$N_{i(k,1)}$ - first hole number for the kth WR (standard $N_{i(k,1)} = 1$)

$N_{i(k,2)}$ - hole number for the kth WR as shown in Fig. 2

$N_{i(k,3)}$ - last hole number for the kth WR as shown in Fig. 2 (by default, for all spindles except the last one, $N_{i(k,3)} = LO$)

$X_{i(k,1)}$ - coordinate in the X axis of the first hole for the kth WR, as shown in Fig. 2

$X_{i(k,2)}$ - coordinates of the hole in the X axis of the kth WR, as shown in Fig. 2.

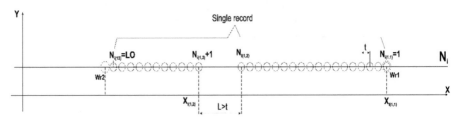

Fig. 3. Symbols used to describe the parameters of a single N_i record WR_1; $N_{1(1,1)}$, $N_{i(1,2)}$, $N_{i(1,3)}$ – number of holes manufactures by spindle Wr1

3   Hole Drilling Algorithm

The basic criterion for making all holes is to choose a drilling strategy for which the drilling time is the shortest. This criterion can be modified slightly and can be reduced to the following wording: maximize the utilization rates of all available drilling spindles. At best, this will be 100% when all spindles make the same number of LO holes, but in plenty of cases, depending on the number of holes in a single row, it will be less than 100%.

To optimize the execution of all holes in a given N_i row, you will find the number of LO holes that each WR spindle (except the last one) will make, for which the utilization rate of all SWWR spindles will be the highest.

The degree of utilization of SWWR spindles will be assessed on the basis of the formula:

$$SWWR = ((L_{WR} - 1) \times LO + R)/(L_{WR} \times LO) \tag{3}$$

where:

SWWR - degree of utilization of available spindles,

L_{WR} - number of available spindles,

R - the number of holes made by the last spindle (the so-called Remainder).

In special cases where $N_i = L_{WR} \times LO$, i.e. when the number of N_i holes can be completely divided by L_{WR}, the utilization rate of SWWR spindles will reach 100%.

Otherwise, if the rest of R remains, with dividing the number of N_i holes by the number of L_{WR} spindles, the degree of use of SWWR spindles will be less than 100%.

Due to the construction of the machine tool and the resulting limitation of displacements in the X axis of individual spindles, there is a certain minimum number of LO_{min} holes, below which each of the spindles cannot make $LO < LO_{min}$ number of holes. This limitation is due to the fact that a minimum distance of L between two adjacent spindles is a design feature of a machine tool and is designated as L_{min}. Therefore, the minimum number of LO_{min} holes meet the condition:

$$LO_{min} = INT \left(L_{min}/t \right) + 1 \tag{4}$$

where:

INT - the integer part of a real number,

L_{min} - minimum distance between two adjacent spindles.

The same applies to the limitation of the reference to the maximum number of LO_{max} holes, which results from the maximum spacing of two adjacent L_{max} spindles, i.e.

$$LO_{max} = INT (L_{max}/t) + 1 \tag{5}$$

The optimum number of LO_{opt} holes that can be made by individual spindles must meet the following condition

$$INT \left(L_{min}/t + 1 \right) <= LO_{opt} <= INT (L_{max}/t + 1) \tag{6}$$

The limitations and criteria for LO_{opt} exploration outlined above are valid if the number of holes in a single N_i row satisfies the following conditions

$$\text{Integer, rounded off} \left(N_i/LO_{min} \right) >= L_{WR} \tag{7}$$

The above condition means that all spindles will be active, i.e. they will take part in the N_i hole construction.

If this condition is not met, it means that fewer spindles than it results from the design capabilities of the L_{WR} machine tool will be used for making N_i holes. This in turn means that the utilization rate of all SWWR spindles will be less than or equal to at least

$$SWWR < (L_{WR} - 1)/L_{WR} \tag{8}$$

where: L_{WR} - number of available spindles on the machine tool.

4 Optimization Algorithm

It has been assumed that the function of the target $f(n)$ of N_i holes on a machine tool equipped with L_{WR} spindles will be a function showing how the number of holes of the rest of R, made by the last spindle differs from the number of LO holes made by all the other spindles, i.e.: the number of holes of the rest of R, made by the last spindle, and the number of LO holes made by all the other spindles:

$$f(n) = (LO - R) \tag{9}$$

where:

   $f(n)$ - function of the target,
   R - number of holes made by the last spindle,
   n – auxiliary variable (natural number 1,2,3… n_{max}), that allows a systematic search.

The minimum value of the target function $f(n)$ will be searched for, which the rest of the R of the holes on the last spindle differs from the number of LO holes made by all other spindles to the smallest extent. In the best case, the function $f(n)$ will take the value 0 (zero), i.e. when the rest of R = LO. Therefore, the number of LO holes will be an independent variable, the optimum LO_{opt} value of which will provide a minimum value of $f(n)$.

There is a minimum number of holes in a single row $N_{i(min)}$ for which all available L_{WR} spindles must be used, namely one for which identity is fulfilled:

$$L_{WR} = \text{Integer Rounded off} \left(N_{i(min)}/LO \right) \tag{10}$$

where: $N_{i(min)}$ - minimum number of holes required in a single row, for which it is necessary to use all available L_{WR} spindles.

If the number of holes N_i in a single row is less than $N_{i(min)}$ then some of all available spindles will not be used, which will significantly reduce the degree of use of SWWR spindles. Then the optimization algorithm will be similar to the $N_i > N_{i(min)}$ case, but instead of the available number of L_{WR} spindles we will need to take the $L_{WR(min)} < L_{WR}$.

Continuing with the optimization procedure, in the first step we check the condition

$$R(N_i/L_{WR}) = 0 \tag{11}$$

where: R – remainder from the division.

In this case, all spindles will be fully utilized and SWWR = 100%. The optimization procedure is interrupted.

If the remainder of the R division is greater than zero, the optimization procedure must be continued.

Then, in order to formulate the optimization algorithm, we introduce the auxiliary concept of the number of holes LO_{aux} in the form of

$$LO_{aux} = \text{INT} \left(N_i/(L_{WR} - 1) \right) \tag{12}$$

where: LO_{aux} - an auxiliary quantity, showing how many LO holes each spindle could potentially make, if only $(L_{WR} - 1)$ spindles were taken into account.

In this case, the minimum remainder of the holes R_{min}, which would be made by the last spindle, is determined as follows:

$$R_{min} = N_i - (L_{WR} - 1) \times LO_{aux} \tag{13}$$

The number of holes LO made by a single spindle can be shown as follows:

$$LO = LO_{aux} - (n - 1) \tag{14}$$

where n - natural number 1, 2, 3....

This dependence was derived from the following reasoning: from the number of holes LO_{aux}, which corresponds to the number of holes made by each of the $(L_{WR} - 1)$ spindles, we subtract $(n - 1)$ holes, and assign $(n - 1)(L_{WR} - 1)$ holes to the last spindle. By changing n from n = 1, then 2, 3 ... we can find its value for which the number of holes made by the last spindle will come close to the number of LO. This will be the optimal solution [5].

In this case, the remainder of the R-holes made by the last spindle can be determined as:

$$R = N_i - (L_{WR} - 1) \times LO \tag{15}$$

and the function of the target f(n) will take the form of

$$f(n) = LO - N_i + (L_{WR} - 1) \times LO = L_{WR} \times LO - N_i \tag{16}$$

$$f(n) = L_{WR} \times (LO_{aux} - (n - 1)) - N_i \tag{17}$$

$$f(n) = L_{WR} \times (INT(N_i/(L_{WR} - 1)) - (n - 1)) - N_i \tag{18}$$

The condition for the optimization of the target function f(n) takes the form of

$$f(n) = (LO - R) = L_{WR} \times (INT(N_i/(L_{WR} - 1) - (n - 1)) - N_i \rightarrow minimum \quad thus\, n = n_{opt} \tag{19}$$

The minimum value of the function f(n) corresponds to the optimum natural number of n_{opt}, which corresponds to the optimum number of LO_{opt} holes.

To illustrate the functioning of the optimization procedure, the following example will be considered:

$t = 20$ mm
$N_i = 101$
$L_{min} = 300$ mm
$L_{WR} = 5$

The values of the following parameters have been determined for the received input data:

$$LO_{min} = INT(300/20) + 1 = 16 \tag{20}$$

$$LO_{aux} = INT(101/(5 - 1)) = 25 \tag{21}$$

$$R_{min} = 101 - (5 - 1) \times 25 = 1 \tag{22}$$

and the function of the target will take the form of

$$f(n) = L_{WR} \times (INT(N_i/(L_{WR} - 1)) - (n - 1)) - N_i = 5 \times (INT(101/(5 - 1)) - (n - 1)) - 101 \tag{23}$$

We need to find the natural number n for which f(n) reaches its minimum.

Table 1a presents the results of the simulation, the effect of the number n on the value of the target function f(n), the number of holes LO and the remainder R for Ni = 101 holes, and Table 1b shows the simulation results for the number of N_i holes, from 16 to 165.

Figure 4 shows the effect of the number of holes N_i in one row on the optimum number of LO_{opt} holes made by a single spindle and the degree of use of available SWWR spindles. This simulation concerns the graduation t = 20 mm and the minimum distance $L_{min} = 300$ mm.

The simulation results presented in Fig. 4 show that depending on the number of holes N_i, the degree of use of available SWWR spindles is in the range of 20–100%.

The results of simulation optimization studies concern cases where the number of holes in the row $N_i \geq 73$. Because then the condition $LO \geq LO_{min} = 16$ is fulfilled, which is the condition for the existence of an optimal solution.

If the number of holes N_i is lower than 73, the best solution resulting from the LO_{min} limitation is

$$LO = LO_{min} = 16 \tag{24}$$

and the last spindle will do the remainder of the R holes, which will be

$$R = N_i - (L_{WR} - 1) \times LO_{min} \tag{25}$$

The degree of utilization of available SWWR spindles will be as follows

$$SWWR = ((L_{WR} - 1) \times LO_{min} + R))/(L_{WR} \times LO_{min}) \tag{26}$$

In Fig. 4 this applies only to the corresponding points $N_i = 73$ and $N_i = 68$ holes. Then $LO = LO_{min} = 16$ and the number of needed spindles is equal to the number of available spindles (in this example $L_{WR} = 5$).

In turn, for the number of holes N_i the number of $L_{WR(min)}$ spindles needed is smaller than the number of available L_{WR} spindles (in this example 5) and amounts to

$$L_{WR(min)} = \text{Integer Rounder off} (N_i/LO_{min}) = 4 < 5 \tag{27}$$

Then the modification of the optimization algorithm is based on the fact that in place of $L_{WR} = 5$ we insert one spindle less (in this example 4). The optimum number of holes, resulting from the limitation of L_{min}, by $(L_{WR} - 1)$ spindles is LO_{min} (in this example 16). In Fig. 4 this refers to the number of holes $N_i = 64, 62$ and 51.

In turn, for an even lower number of holes $N_i \leq 48$ we need to reduce the number of necessary spindles by one (to 3 in this example) and use the proposed algorithm. In Fig. 4 this applies to the number of holes $N_i = 48$ and 33.

Optimization of the Drilling Process in Multi-hole Objects   131

Table 1. Results of the simulation of the optimization procedure: a) for $N_i = 101$, b) for $N_i = 16$ – 165

a)

n	f(n)	LO	R(n)
1	24	25	1
2	19	24	5
3	14	23	9
4	9	22	13
5	4	21	17
6	-1	20	21
7	-6	19	25
8	-11	18	29
9	-16	17	33
10	-21	16	37
11	-26	15	41
12	-31	14	45
13	-36	13	49
14	-41	12	53
15	-46	11	57
16	-51	10	61
17	-56	9	65
18	-61	8	69
19	-66	7	73
20	-71	6	77
21	-76	5	81

b)

N	f(n)$_{min}$	LO	R	LO>LO$_{min}$	SWWR
165	0	33	33	TRUE	100,0
160	0	32	32	TRUE	100,0
155	0	31	31	TRUE	100,0
150	0	30	30	TRUE	100,0
141	4	29	25	TRUE	97,2
133	2	27	25	TRUE	98,5
123	2	25	23	TRUE	98,4
113	2	23	21	TRUE	98,3
101	4	21	17	TRUE	96,2
90	0	18	18	TRUE	100,0
87	3	18	15	TRUE	96,7
81	4	17	13	TRUE	95,3
73	2	16	9	FALSE	91,3
68	2	16	4	FALSE	85,0
64	1	16	0	FALSE	80,0
62	3	16	14	FALSE	77,5
51	4	16	3	FALSE	63,8
48	2	16	0	FALSE	60,0
33	2	16	1	FALSE	41,3
20	0	16	4	FALSE	25,0
16	4	16	0	FALSE	20,0

In Fig. 4 you can see that starting with $N_i <= 73$ $LO_{opt} = LO_{min}$, (in this example 16). While for more than 73 holes in a single row LO_{opt} increases.

Figure 4 also shows the effect of the number of holes N_i on the degree of use of available SWWR spindles. For $N_i > = 81$ (this is the minimum number of holes for which LO_{opt} is greater than LO_{min}) SWWR is within a range of (95–100)%. Whereas for $N_i < 81$ holes, SWWR decreases as far as 20% (for $N_i = 16$).

In order to develop a program for a multi-spindle CNC drilling machine, geometric information about the X coordinate of each hole is necessary. The $X_i(k, j)$ coordinate, i.e. the coordinate of the jth hole in the ith row made by the kth WR spindle can be determined according to the following rule:

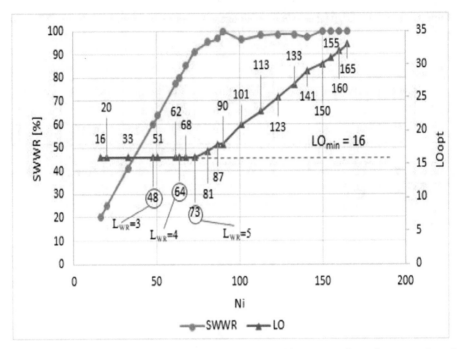

Fig. 4. Influence of the number of holes N_i in a single partition row on the optimum number of LO_{opt} holes and the utilization rate of SWWR spindles for $t = 20$ mm

a) we determine the kth spindle, which must satisfy the unevenness

$$(k - 1) <= \text{Integer Rounded off } (j/LO_{opt}) <= k \qquad (28)$$

hence the spindle number k.

b) we determine the $X_{i(k,j)}$ coordinate for the two cases, the first when there is no $L >$ t situation (see Fig. 2 for WR_1) and the second when there is this situation.

The first case occurs when the following identity is fulfilled

$$\text{MODULE}(X_{i(k+1,1)} - X_{i(k,1)}) = (LO_{opt} - 1)\, t \qquad (29)$$

i.e. when the distance between subsequent spindles is equal to a multiple (LO_{opt}) of the scale t (in Fig. 2 it would correspond to the situation when $L = t$). Then the $X_{i,j}$ coordinate is calculated from the formula:

$$X_{i,j} = X_{i(k, j)} = X_{i(k, 1)} + [j - (k - 1)LO_{opt}]\, t \qquad (30)$$

The second case occurs when the following identity is fulfilled

$$\text{MODULE}(X_{i(k+1,1)} - X_{i(k,1)}) > (LO_{opt} - 1)\, t \qquad (31)$$

i.e. when the distance between subsequent spindles is higher than a multiple (LO_{opt}) of the scale t (in Fig. 2 it would correspond to the situation when $L > t$). Then the $X_{i,j}$ coordinates are calculated according to two formulas:

1. when $(k - 1) \, LO_{opt} <= j <= N_{i(k,1)}$ (in Fig. 3, for example, this applies to holes for WR_1 from the first $N_{i(1,1)}$ to $N_{i(1,2)}$), according to the formula

$$X_{i,j} = X_{i(k,\,j)} = X_{i(k,1)} + \left[j - (k - 1)LO_{opt} \right] t \tag{32}$$

2. when $N_{i(k,1)} <= j <= k \, LO_{opt}$ (in Fig. 3, for example, this applies to holes for WR_1 from $N_{i(1,2)} + 1$ to $N_{i(1,3)}$), according to the formula

$$X_{i,j} = X_{i(k,j)} = X_{i(k,2)} + \left(j - N_{i(k,2)} \right) t \tag{33}$$

5 Summary

The paper discusses the machining of multiple holes in large-sized objects on a multi-spindle CNC machine tool (drilling machine). Regarding the very long time for development of such an object (a dozen to several dozens of hours) there is a need to optimize such technological operations. The optimization comes down to choosing the holes to be made in such a way as to make the most of all available spindles in the machine tool. This will mean the shortest possible time for all holes to be made. It is therefore a discreet optimization problem [5, 6]. The paper introduces the spindle use indicator (degree) SWWR and proposes a discrete function of a purpose, which represents the difference between the number of holes made by all L_{WR} spindles despite the last one, and the so called R Remainder, namely the number of holes made by the last spindle. For this function of the target we search for a natural number n, for which it takes a minimum value. This natural number allows to calculate the number of holes made by each spindle LO_{opt}, except for the last one, and this is the optimal solution. The paper presents numerical examples of such optimization.

References

1. Pudlik, W.: Wymiana i wymienniki ciepła. Polish Scientific Publisher PWN, Warsaw (2012)
2. Collective work.: Obrabiarki do skrawania metali. Scientific and Technical Publishing WNT, Warsaw (1974)
3. Pająk, E., Wieczorowski, K.: Podstawy optymalizacji operacji technologicznych w przykładach. Polish Scientific Publisher PWN, Warsaw (1982)
4. Jarosz, K., Löschner, P., Królczyk, G.: Weryfikacja i optymalizacja programów sterujących na frezarki CNC metodą elementów skończonych w środowisku programu TWS Production Module 3D. Mechanik (2016)
5. Danielewska-Tułecka, A., Kusiak, J., Oprocha, P.: Optymalizacja. Polish Scientific Publisher PWN, Warsaw (2009)
6. Wittbrod, P.: Optymalizacja procesu obróbki skrawaniem z wykorzystaniem algorytmów genetycznych. www.ptzp.org.pl

Comparative Tests of Static Properties of Steel and Polymer Concrete Machine Tool Body Elements

Jan Kosmol$^{(\boxtimes)}$ ⓘ, Krzysztof Lis ⓘ, and Paweł Całka ⓘ

Silesian University, 44-100 Gliwice, Poland
jan.kosmol@polsl.pl

Abstract. The article presents preliminary results of experimental research on an exemplary steel machine tool body support beam unfilled and filled with polymer concrete and from homogeneous polymer concrete. The scope of the research involved an attempt to determine the static properties of the beam depending on its design. The tests were carried out for a very small range of forces due to the danger of damage to beams made of polymer concrete alone. The same test conditions were used for all types of beams. As a result of the tests, the deflection characteristics of a double-sided beam were obtained. During the research, the configuration of the measuring apparatus and dedicated software for testing and analysing the results were developed. An induction displacement sensor with measurement accuracy up to 0.5 μm and a strain gauge force sensor with an accuracy of 0.2 N was used in the station. The force actuator was an electrodynamic inductor controlled by a microprocessor system controlling the given force. Each measurement was repeated ten times. The test results were used to determine parameters such as the stiffness coefficient. Due to the fact that the beams were not symmetrical, the tests were carried out for the beam in two directions. The obtained characteristics allowed the conclusions about the comparison of the used beams and they are a supplement to the research conducted by the authors regarding their dynamic properties.

Keywords: Machine tool body · Polymer concrete · Static properties

1 Introduction

The continuous development of machining technology forces machine builders to search for better, more accurate and more efficient construction solutions. In machine tools, one of the most important elements of the whole machine is its body because it is the body that the final machining capabilities of the machine depend largely on. The main purpose of the body is to seal all the other mechanisms of the machine but also to give it adequate rigidity. The most popular are cast iron bodies, inter alia due to good dynamic properties, relatively cheap production and the possibility of making more complex shapes when compared to steel bodies. Welded steel structures are characterized by high rigidity, however, their dynamic properties, in particular the ability to damp the vibrations arising during operation is very small [1, 2].

© The Author(s), under exclusive license to Springer Nature Switzerland AG 2021
A. Mężyk et al. (Eds.): SMWM 2020, AISC 1336, pp. 134–144, 2021.
https://doi.org/10.1007/978-3-030-68455-6_12

In recent years, a growing interest in hybrid and composite machine tool bodies can be noticed. Some machine tool manufacturers are departing from classic cast iron bodies, replacing them with bodies made of composite materials. The main reason for these changes are very good dynamic properties of composites and better thermal properties compared to cast iron bodies. The material most commonly used in the production of composite bodies is polymer concrete. Polymer concrete, also known as resin concrete, is a mixture of aggregates with different gradations together with a two-component synthetic resin. A properly selected and made mixture allows to obtain coherent castings, characterized by very high vibration damping, low heat transfer coefficient and relatively good rigidity [5–19]. The material data for polymer concrete compared to steel and cast iron are presented in Table 1.

The tests carried out so far have confirmed that the polymer concrete has very good dynamic properties. Figure 1 presents examples of dimensionless coefficient of damping obtained during the testing of dynamic properties for a steel profile, polymer concrete casting and a steel profile filled with polymer concrete. The obtained values of the coefficient refer to the first three bending forms of vibrations [3, 4].

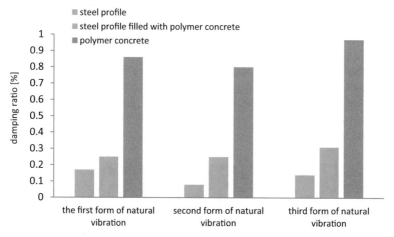

Fig. 1. Comparison of vibration damping coefficients for steel, steel filled with polymer concrete (hybrid combination) and polymer concrete casting [3, 4].

However, apart from good vibration damping capability, the body must also have adequate rigidity. Hard materials machining requires the use of appropriate cutting parameters which generate significant forces during machining. When the rigidity of the body is too low, the resulting forces can cause too much displacement of its individual elements, which may affect the accuracy of the manufactured detail. Therefore, the next step in the research on the properties of polymer concrete was to determine the rigidity of the composite samples made, comparing them to the rigidity of the steel profile and hybrid connection.

Table 1. Material data [10].

Properties	Polymer concrete	Steel	Cast iron
Young module [kN/mm^2]	30–40	210	80–120
Tensile strength [N/mm^2]	13–15	300	200
Compressive strength [N/mm^2]	140–170	250–1200	350–450
Density [g/cm^3]	2.1–2.4	7.85	7.2
Logarithmic damping decrement	0.02–0.03	0.002	0.003
Poisson factor	0.25	0.3	0.26

2   Research Object

An empty steel profile with dimensions of $120 \times 60 \times 3$ was subjected to rigidity tests, all the beams made had the length of 1400 mm. The other tested object was a steel profile filled with polymer concrete. The casting used in this case consisted of 10% epoxy resin and a mixture of quartz aggregates. Resin castings which differed from each other in their material composition were also made and tested. One beam was made of an identical mixture that filled the steel profile. The other beam was made of an identical aggregate mix but using a 20% resin content. Literature resources allow to conclude that the optimal resin content should be in the range of 10–20%. Too little amount of resin in the casting does not allow its proper fusion and negatively affects the final properties. On the contrary, castings with too much resin content are prone to uneven aggregate distribution, which also negatively affects its properties. The tested samples are presented in Fig. 2 to 5.

Fig. 2. Beams tested for stiffness properties - steel profile filled with polymer concrete.

Fig. 3. Beams tested for stiffness properties - steel profile.

Fig. 4. Beams tested for stiffness properties - polymer concrete 10% of resins.

2.1 Test Stand

The experimental static stiffness test was carried out with vertical and horizontal positioning of the beams. Each beam was supported at three points, closest to the edge. The scheme of the measuring track is presented in Fig. 8. An inductor was placed in the centre of the beam perpendicular to the top surface, which generated the force causing the beam to deflect. During the tests, the Modal Shop TMS 2007 electrodynamic force generator (Fig. 6) with a modified range up to 70 N was used, while the force with which the inductor acted on the beam was measured using a Mark MR4–10 strain gauge force sensor with a range of up to 100 N and accuracy of ±0.1 N with a Mark 5i conditioner (Fig. 7).

Fig. 5. Beams tested for stiffness properties - polymer concrete 20% of resins.

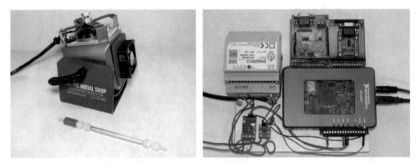

Fig. 6. Modified Modal Shop TMS 2007 electrodynamic force generator with cooling system and control system based on National Instrument MyRIO.

Fig. 7. Mark MR4-10 strain gauge force sensor with a Mark 5i conditioner and Sylvac D300S conditioner with P5 displacement sensor.

The inductor had a control system built for the purpose of testing enabling the implementation of constant force. The current regulator in the inductor winding was implemented into a microprocessor system that also performed the function of data acquisition and controlling the experiment. The data in the form of developed characteristics were sent to the NI PXI computer (host), which had the experiment control panel. The displacement sensor Sylvac P5 with an accuracy of 0.5 μm together with the D300S conditioner was responsible for the displacement measurement (Fig. 7). The displacement and force measurement system is equipped with digital communication interfaces sending data to the microprocessor system (National Instrument MyRIO).

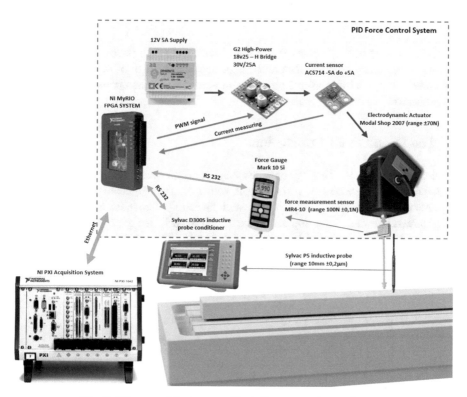

Fig. 8. Diagram of the test stand including measuring equipment.

The test stand is shown in Fig. 8. The base of the beams was a monolithic cast iron table of a heavy conventional machine tool. Additionally, mechanical dial indicators with an accuracy of 1μm were used in the support points. The sensors were used to verify the correctness of beam support. During the experiment no change in their indications was observed.

Each beam was tested in 10 measuring series. In each series, the interacting force increased to the value of several dozen Newtons, then decreased back to zero. The value of the maximum force causing deflection of the beam was due to the non-destructive nature of the tests (Fig. 9).

Fig. 9. Test stand with sample beam - horizontal configuration.

When testing beams with expected high bending strength (steel, filled steel) much greater forces could be used. Due to the expected low strength of monolithic polymer concrete beams, the impact force on the beam was limited to a maximum value of 50 N for all beams.

3 Test Results and Conclusions

As an example, Fig. 10 shows the obtained average displacement values for a beam made of polymer concrete with 10% resin content, with the horizontal beam position. As it can be observed, the relation of displacement and the acting force is linear. A certain value of hysteresis between increasing and decreasing force can also be seen. The graphs also have error bars $\pm 1\sigma$ for individual measurement series. The error is very small and does not exceed 0.5 μm for all tests.

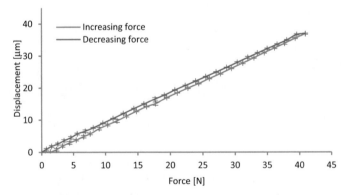

Fig. 10. Example of deflection characteristics as a function of epoxy polymer concrete 10% of resins for vertically configuration.

The characteristics presented in Fig. 11 refer to the beam in the form of a steel profile. It can be noticed that the hysteresis in this case is lower than in the case of a beam filled with polymer concrete (Fig. 12) despite twice the rigidity of such a beam. The value of this hysteresis does not exceed the estimated measurement error, i.e. 0.5 μm. For a beam filled with polymer concrete, the hysteresis is much higher, i.e. 1.5 μm.

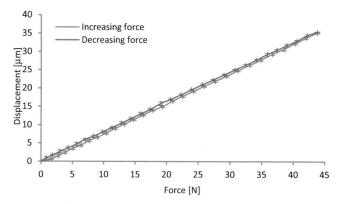

Fig. 11. Example of deflection characteristics as a function of steel profile for vertically configuration.

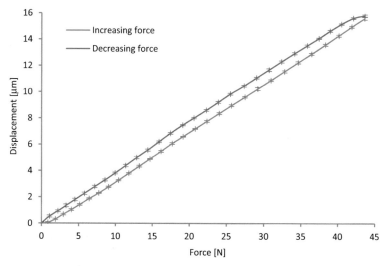

Fig. 12. Example of deflection characteristics as a function of force for an steel profile filled with polymer concrete for vertically configuration.

Tables 2 and 3 summarize the results of the experiment and present the basic statistics. The steel profile filled with polymer concrete shows the highest stiffness.

Table 2. Results of tested beams for vertically configuration.

Subject of study	Max. Force [N]	Max. deflections [μm]	Stiffness [N/μm]	Std. dev. of stiffness [N/μm]
Steel profile	43.8	35.4	1.2	0.005
Steel profile filled with Polymer concrete	44.2	15.9	2.7	0.018
Polymer concrete 10% of resins	41.3	37	1.1	0.004
Polymer concrete 20% of resins	37.8	49.6	0.8	0.002

Table 3. Results of tested beams for horizontal configuration.

Subject of study	Max. Force [N]	Max. deflections [μm]	Stiffness [N/μm]	Std. dev. of stiffness [N/μm]
Steel profile	42.6	12.4	3.4	0.022
Steel profile filled with polymer concrete	41.9	6.2	6.8	0.081
Polymer concrete 10% of resins	42.3	11.2	3.8	0.040
Polymer concrete 20% of resins	40.3	15.5	2.6	0.015

When the beam is positioned horizontally, its rigidity is more than twice as high as the steel profile. These results show that filling the steel profile not only positively affects its ability to damp vibrations, raising it several times in relation to the hollow profile as proved in other studies of the authors, among others in [3], but it also significantly increases its rigidity. However, it must be taken into consideration that this is also the beam with the largest mass. It can also be seen that the rigidity of polymer concrete with a 10% resin content is very similar to an empty steel profile, which is a very interesting conclusion, especially in the context of a direct replacement of steel structures with resin ones. The problem is the tensile strength of polymer concrete castings. This shows that the polymer concrete casting not only has great vibration damping properties but its rigidity can be comparable to a steel profile. The casting with 20% resin content has the worst rigidity. From a technological point of view, a larger amount of resin allows to mix and fill the mould more easily, but it can be observed that with this amount of resin, the casting properties are worse than with 10% resin.

It would also be necessary to identify a large displacement hysteresis value for a steel beam filled with polymer concrete. Despite the good dynamic and static properties confirmed in this study, the stability of geometrical features of constructions made in body casting technology must be verified. The hysteresis calls into question the repeatability of positioning of machines made in this technology in favour of machines in which the bodies were made as a monolithic cast. The problem may arise from adhesive forces maintaining the integrity of the filling and steel. When there is a loss of consistency, friction appears that subsequently generates hysteresis. This phenomenon may depend on the temperature and the way the body is used (loading, corrosion of steel).

References

1. Bruni, C., Forcellese, A., Gabrielli, F.M., Simoncini, M.: Hard turning of an alloy steel on a machine tool with a polimer concrete bed. J. Mater. Process. Technol. **202**(1–3), 493–499 (2008)
2. Bignozzi, M.C., Saccani, A., Sandrolini, F.: New polimer mortars containing polymeric wastes-part 2: dynamic mechanical and dielectric behaviour. Composites A **33**(2), 205–211 (2002)
3. Całka, P., Lis, K.: Badanie właściwości dynamicznych polimerobetonu w kontekście hybrydowych korpusów obrabiarek. Inżynieria Maszyn, R. 23, z, 1, str. 77–86 (2018)
4. Całka, P., Lis, K.: Eksperymentalna analiza modalna profilu stalowego wypełnionego polimerobetonem. STAL Metale & Nowe Technologie, nr 9–10, str. 33-37 (2017)
5. Chmielewska, B., Czarnecki, L., Sustersic, J., Zajc, A.: The influence of silane coupling agents on the polymer mortar. Cem. Concr. Compos. **28**(9), 803–810 (2006)
6. Cort´es, F., Castillo, G.: Comparison between the dynamical properties of polymer concrete and grey cast iron for machine tool applications. Mater. Des. **28**(5), 1461–1466 (2007)
7. Dolinšek, S., Šuštaršič, B., Kopač, J.: Wear mechanisms of cutting tools in high-speed cutting processes. Wear **250**, 349–356 (2001)
8. Erbe, T., Król, J., Theska, R.: Mineral casting as material for machine base-frames of precision machines. In: Twenty-third Annual Meeting of the American Society for Precision Engineering and Twelfth ICPE, Portland, Oregon (2008)
9. Gorninski, J.P., Dal Molin, D.C., Kazmierczak, D.C.: Strength degradation of polymer concrete in acidic environments. Cem. Concr. Compos. **29**(8), 637–645 (2007)
10. Kosmol, J.: Zastosowanie polimerów w budowie korpusów obrabiarek. Przetwórstwo tworzyw, nr 1, str. 33–45 (2017)
11. Kosmol, J., Wilk, P.: Próba optymalizacji korpusu obrabiarki z zastosowaniem MES i algorytmu genetycznego. Modelowanie inżynierskie, nr 35, str. 59–66 (2008)
12. Kosmol, J., Lis, K., Całka, P.: Eksperymentalna weryfikacja korpusów polimerobetonowych. Przetwórstwo tworzyw, nr 1, str. 39–47 (2018)
13. Dos Reis, J.M.L.: Mechanical characterization of fiber reinforced polymer Concrete. Mater. Res. **8**(3), 357–360 (2005)
14. Muthukumar, M., Mohan, D., Rajendran, M.: Optimization of mix proportions of mineral aggregates using box Behnken design of experiments. Cem. Concr. Compos. **25**(7), 751–758 (2003)
15. Orak, S.: Investigation of vibration damping on polymer concreto with polyester resin. Cem. Concr. Res. **30**(2), 171–174 (2000)
16. Ribeiro, M., Tavares, C.M.L., Figueiredo, M., Ferreira, A.J.M., Fernandes, A.A.: Bending characteristics of resin concretes. Mater. Res. **6**(2), 247–254 (2003)

17. Ribeiro, M.C.S., N'ovoa, P.R., Ferreira, A.J.M., Marques, A.T.: Flexural performance of polyester and epoxy polymermortars under severe thermal conditions. Cement and Concrete Composites, vol. 26, no. 7, pp. 803–809 (2004)
18. Tawfik, M.E., Eskander, S.B.: Polymer concrete from marble wastes and recycled poly (ethylene terephthalate). J. Elastomers Plast. **38**(1), 65–79 (2006)
19. Xu, P., Yu, Y.H.: Research on steel-fibber polymer concreto machine tool structure. J. Coal Sci. Eng. **14**(4), 689–692 (2008)

The Use of Beta Statistical Distribution in Speed Modeling Of Military High Mobility Wheeled Vehicle In Proving Ground Tests

Mariusz Kosobudzki[(✉)] [ID]

Wroclaw University of Science and Technology, Łukasiewicza Str. 7/9, 50-371 Wrocław, Poland
mariusz.kosobudzki@pwr.edu.pl

Abstract. Qualifying tests of new vehicles are aimed at showing that the manufactured one meets the requirements of the user and are carried out by authorized research institutions, most often as proving ground tests. In such tests, it is important to make appropriate initial assumptions, including those regarding the speed of the vehicle being tested. Determining vehicle speed during tests has a direct impact on recorded loads, which shapes the results obtained. The article presents the use of the beta statistical distribution to determine vehicle speed. It allows shaping the driving speed in such a way that the maximum and average values in a given test meet the initial assumptions resulting from the operation profile.

Keywords: Proving ground tests · High mobility multipurpose wheeled vehicle · Driving speed modelling

1 Introduction

High mobility multipurpose wheeled vehicles are the primary means of transport for the subunits of the Armed Forces. They are usually constructed as a universal chassis adapted to fit a transport box with a canvas cover, a standardized container with special equipment or permanent special equipment [20]. Low probability of the operational failure occurrence, usually below 5%, and the required durability ought to characterize them [8]. According to the developed concept of operations, they are planned to be used in various road conditions and off-road, day and night, all year round, with an intensity of up to 24 h a day. These vehicles must be adapted to cooperate with combat vehicles on both wheeled and tracked chassis in terms of their driving speeds (maximum and average) and their mobility indicators. The doctrinal documents assume that the daily driving time of a vehicle may amount to 12 h [22, 24].

High mobility wheeled vehicles with special bodies are used primarily in the logistics operations. A mobile workshop, which technically supports for combat equipment, is an excellent example of such a vehicle. The workshop crew operates in the area behind the combat subunits at a distance of around $1 \div 3$ km and is called upon to provide technical assistance that should not last longer than 30 min, and the whole operation - not longer than 1 h. In special situations, a damaged vehicle may be evacuated by towing to the area where technical subunits repairing damaged military equipment operate. The average movement speed in cross-country conditions should, therefore, be about 4 ÷ 5 km/h. With due regard to the crew's operating time (10 ÷ 12 h per day), such a vehicle can cover about 50 ÷ 60 km per day.

Another example of using the same vehicle is tactical deployment, which takes place on unpaved and hardened roads. The dislocation distance is from a dozen to several dozen kilometers. Under the assumption that the relocation should be executed within 1 ÷ 2 h, the average speeds achieved will be of about 20 ÷ 25 km/h.

One more example is the technical support for the marching column (distances over 50 km), where the movement of vehicles takes place on hardened roads. The average vehicle speeds (depending on the types of vehicles included in the marching column) may be 25 ÷ 50 km/h. In the case of the vehicle's independent drive (usually on hardened roads), the average speeds achieved are about 55 ÷ 60 km/h [17, 21].

The examples below assume that the vehicle will run under varied road conditions at different average and maximum speeds on different types of roads and off-roads. Table 1 presents the summary of the expected conditions of the vehicle movement limited to four characteristic road surfaces.

Table 1. Expected conditions of the vehicle movement.

No.	Surface type	Estimated average speed [km/h]	Estimated maximum speed [km/h]	Maximum possible daily mileage [km]
1	Asphalt, concrete (e.g. motorways, expressways) - good surface quality	60	90	720
2	Asphalt, concrete, cobblestone (e.g. district roads, commune roads) - medium and low surface quality	50	80	600
3	Off-roads	25	40	300
4	Cross-country	5	10	60

The maximum possible daily mileage of the vehicle shown in Table 1 corresponds to a situation where the driver's daily readiness to work would be used without interruption. The established mileage of high mobility wheeled vehicles of small and medium load capacity ranges to 400 thousand km [4].

2   Vehicle Usage Model for Qualification Tests

Before the decision is made to acquire a vehicle prototype for the Armed Forces, it undergoes qualification tests organized and conducted by a designated testing or certification body [14]. Within the framework of this procedure, a test program concerning the confirmation of required vehicle functionalities is agreed with its manufacturer. The most frequently, normal and contractual usage conditions are modelled with the use of test methods such as profile ground and bench tests [9].

Standards are the basis for modelling the conditions of use for profile ground tests concerning the vehicle durability, as an evaluation of the reliability components [6, 7, 13, 16]. The usage model contains, among others, information about the types of surfaces on which the vehicle will run and the share of these road surfaces in the established mileage of the vehicle. It is assumed that the vehicle will be loaded with the nominal load (the total vehicle loading capacity is utilized) during the tests. However, the method for determining the vehicle speed to the identified surface types, which is a key controlled vehicle movement parameter in profile ground testing, is questionable.

In real traffic conditions, vehicle speeds vary from zero to the maximum value that can be determined based on the adopted assumptions included in the concept of operational vehicle usage, restrictions resulting from road traffic regulations [20] and those related to vehicle traffic safety [12, 19] or driver comfort [3, 11, 16, 18].

A Life Cycle Environmental Profile, which indicates the average and maximum speeds at which a vehicle will drive on the identified road surfaces, is developed following the technical standards for military equipment in force at NATO [1]. The data are supplemented with the share of the indicated types of road surfaces in the vehicle target mileage.

Standards [13, 16] directly specify the share of identified four surface types in the established mileage of the tested vehicle. The criterion of dynamic loads acting on the specified elements of the driving system or on the crew members (characteristics of vibrations generated while driving on the driver and crew members and exceeding the level of comfort/nuisance/harmfulness) is applied. The criterion for determining the maximum vehicle speed shall be the permissible dynamic loads acting on the specified parts of the chassis or crew members. Such an approach is different from NATO standards [1]. It requires, among others, pre-test identification of elements of the vehicle chassis that will be subject to assessment, and determining their, e.g., fatigue strength characteristics. All that because this type of strength is crucial in the durability tests of this element [5, 15]. Moreover, if tests are conducted on non-parameterized roads, the results obtained are not representative of the conditions for vehicles' usage in other military units. The method employed in the standard for determining the driving speed during tests [16] is discussed a well. The criterion adopted in the standard is the level of vibrations affecting the driver and passengers, thus reducing their comfort and increasing inconvenience or harmfulness. The course of the vibration threshold includes the resonant frequencies of human internal organs and can be set up from the standard [2]. However, the current national health protection regulations refer today to another indicator characterizing the working environment, which is the value of daily or short-term exposure to vibrations and the values of action thresholds for daily exposure to mechanical vibrations [18].

Nonetheless, regardless of the criterion used to determine the dynamic loading spectrum acting on the selected component or a crew member, representative vehicle speed distributions shall be established. The above should consider the possible speed values allowing for achieving the expected average and maximum speed rates corresponding to the concept of operation vehicle use.

3 Vehicle Speed Distributions for Profile Ground Tests

A statistical beta distribution can be used by normalizing the speed variation ranges with respect to the maximum speed in the test to produce a speed spectrum representative of vehicle movement under different road conditions. In this way, the distribution is limited from the top and the bottom [25]. The percentage of the fixed speed range in relation to the maximum speed can be determined from the formula:

$$f(x) = \frac{(\alpha + \beta - 1)!}{(\alpha - 1)!(\beta - 1)!} x^{\alpha-1}(1 - x)^{\beta-1} \tag{1}$$

where: f(x) - percentage of the adopted speed range throughout the entire vehicle speed range, x – the value of the speed range, α, β - parameters of the distribution shape.

The acceptance of the range of speed variations (the magnitude x in the Eq. (1)) in a subset of analyzed speeds is determined by an engineer from the required resolution of the speed spectrum and the achievable and maintainable speed of the analyzed vehicle. This range may be, e.g., $1.5 \div 4$ km/h for low rate driving on dirt roads and, e.g., $5 \div 10$ km/h for driving on other types of surfaces.

Given the vehicle movement input data, e.g. on roads with an asphalt surface where the assumed speed $V_{avg} = 50$ km/h, $V_{max} = 80$ km/h and the share of the test road section length per 1000 km of mileage is 25% (250 km), one can ascertain data to carry out the tests. The above is shown in Table 2.

Table 2. Beta distribution of driving speed to the established road test parameters - surface no.2 in Table 1 (speed change step - resolution of the speed spectrum $\Delta V = 5$ km/h).

Represented speed V_{rep} [km/h]	V_{min} [km/h]	V_{max} [km/h]	V_{avg} [km/h]	Driving time [%]	Driving time [min]	Sectional mileage [%]	Sectional mileage [km]
–	0	5	2,5	0,0	0,1	0,0	0,0
5	2,5	7,5	5	0,1	0,4	0,0	0,0
10	7,5	12,5	10	0,6	1,7	0,1	0,3
15	12,5	17,5	15	1,3	4,0	0,4	1,0
20	17,5	22,5	20	2,4	7,2	1,0	2,4
25	22,5	27,5	25	3,7	11,0	1,8	4,6
30	27,5	32,5	30	5,1	15,2	3,0	7,6
35	32,5	37,5	35	6,6	19,7	4,6	11,5
40	37,5	42,5	40	8,0	24,0	6,4	16,0
45	42,5	47,5	45	9,3	27,9	8,4	20,9
50	47,5	52,5	50	10,3	30,9	10,3	25,8
55	52,5	57,5	55	11,0	32,9	12,1	30,1
60	57,5	62,5	60	11,1	33,3	13,3	33,3
65	62,5	67,5	65	10,5	31,6	13,7	34,2
70	67,5	72,5	70	9,1	27,4	12,8	32,0
75	72,5	77,5	75	6,7	20,0	10,0	25,0
80	77,5	82,5	80	2,8	8,4	4,5	11,2
				99	**296**	**102**	**256**

The data from Table 3 can be taken for the initial adoption of the value of the shape parameter α.

Table 3. Values of the shape parameter α [1].

V_{avg}/V_{max} [%]	10	20	30	40	50	60	70	80	90
α		0,2	0,5	0,75	1,25	2	3	4	6	13

The relationship below can be used to pre-determine the value of the β parameter:

$$\beta = \frac{\alpha \cdot \vartheta_{max}}{\vartheta_{avg}} - \alpha \qquad (2)$$

where: ϑ_{max}, ϑ_{avg} - the required value of maximum and average speed in the road test.

The results presented in Table 2 were optimized for the criterion of driving time (driving speed) by the iterative selection of the values of shape α and β parameters. The total mileage given in the example is higher than the assumed one by 4%, due to the difficulty of carrying out measurements at the lower speed range (minimum speeds).

Diagrams of the percentage distribution of driving time with the set speed and the percentage values of mileage corresponding to these speeds, as shown in Fig. 1, can be drawn up based on the data presented in Table 2.

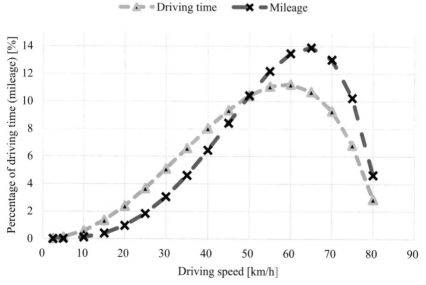

Fig. 1. Percentage beta distribution of driving time and mileage as a function of driving speed $V_{max} = 80$ km/h, $V_{avg} = 50$ km/h

The mileages in Fig. 2 show that, for example in the indicated test, the speed should be maintained at 50 km/h for 30,9 min or the distance of 25,8 km is to be covered at the same speed.

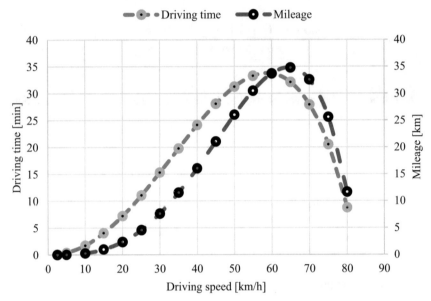

Fig. 2. Driving time and mileage distribution as a function of driving speed $V_{max} = 80$ km/h, $V_{avg} = 50$ km/h.

With predetermined speed distributions, the analysis should be made to confirm whether the assumed vehicle speed ranges on the selected test section are achievable and maintainable during the test.

4   The Analysis of Possible Vehicle Speeds

It is necessary to know the characteristics of the vehicle, including the engine (range of useful rotational speeds of engine crankshaft), the drivetrain (available kinematic ratios) and tires (size, radial stiffness), to analyze the possibility of carrying out road tests in accordance with the adopted speed distributions. Knowledge of the above characteristics allows to indicate whether the assumed driving speeds can be achieved and maintained during the test. The range of useful crankshaft speeds of the engine is assumed to be limited by the presence of the maximum torque from the bottom and the maximum power from the top. Acquaintance with the characteristics of tires allows determining the dynamic radius value and its variation as a function of the air pressure in the tire. According to the requirements of the standard [10], high mobility multipurpose vehicles are equipped with a central tire inflation system, which enables reducing the air pressure by increasing the deflection of the tire. The U tire deflection index is defined according to the relationship:

$$U = \frac{H'}{H} \tag{3}$$

where: H' - the height of the tire loaded nominally at a steady pressure, H - the height of the unladen tire at a nominal pressure.

Table 4 shows selected characteristics of some high mobility wheeled vehicle: the engine, propulsion system, and tires. The given driving speeds V were determined for the average speed of the engine crankshaft (n = 1700 [1/min]) with the range of driving speed change ΔV resulting from the change of the engine crankshaft speed Δn ($\Delta n = \frac{2200-1200}{2} = 500$ [1/min]). The reduction of the driving speed caused by the tire deflection, which occurs after lowering the air pressure, is also considered.

Table 4. Vehicle speed characteristics on the example vehicle.

Max. engine torque: 1300 [Nm]	1200÷1600 [1/min]		Nominal engine power: 240 [kW]	2200 [1/min]		Tire: Cross-country speed correction based on tire deflection:		14.00R20 164/160J		
Road speed [km/h]			**Cross-country speed [km/h]**			U=0,86 (nominal deflection)	U=0,8	U=0,7	U=0,6	U=0,5
Gear	V (n=1700 [1/min])	ΔV (Δn=±500 [1/min])	Gear	V (n=1700 [1/min])	ΔV (Δn=±500 [1/min])	$\Delta V_{0,86}$	$\Delta V_{0,8}$	$\Delta V_{0,7}$	$\Delta V_{0,6}$	$\Delta V_{0,5}$
C	6,9	2,0	C	3,4	1,0	0,0	−0,1	−0,2	−0,3	−0,4
1	9,9	2,9	1	4,9	1,5	0,0	−0,1	−0,3	−0,4	−0,6
2	13,9	4,1	2	6,9	2,0	0,0	−0,2	−0,4	−0,6	−0,8
3	18,7	5,5	3	9,3	2,7	0,0	−0,3	−0,5	−0,8	−1,1
4	24,8	7,3	4	12,4	3,6	0,0	−0,4	−0,7	−1,1	−1,5
5	34,4	10,1	5	17,2	5,1	0,0	−0,5	−1,0	−1,6	−2,1
6	48,2	14,2	6	24,1	7,1	0,0	−0,7	−1,4	−2,2	−2,9
7	65,0	19,1	7	32,5	9,6	0,0	−0,9	−1,9	−2,9	−3,9
8	86,7	25,5	8	43,4	12,8	0,0	−1,2	−2,6	−3,9	−5,3
R	7,2	2,1	R	3,6	1,1	0,0	−0,1	−0,2	−0,3	−0,4

The lowest possible forward speed (gear C in cross-country conditions) is 2.5 km/h and may be reduced by 0.4 km/h if the tire inflation pressure is decreased (U = 0.5). The characteristic of tire deflection caused by the change of inflation pressure is presented in Fig. 3.

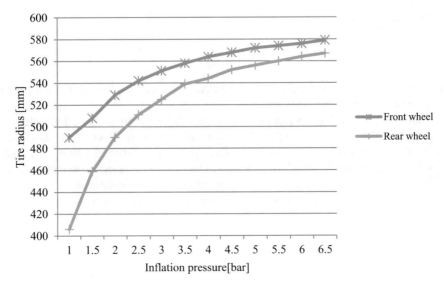

Fig. 3. The characteristic of tire deflection caused by the change of inflation pressure (max deflection is limited by run flat pad).

5   Conclusions

The proposed method for distributing the vehicle speed into a statistically equivalent set of easily measurable physical quantities (driving time at a fixed constant speed or the length of the road measuring distance covered at a set speed) allows specifying the vehicle movement parameters, thereby ensuring their representativeness during the conducted tests. The assumed acceptable range of speed changes in each measurement step could be determined individually to the road conditions, measurement equipment, or limitations of the vehicle (movement speeds that are possible to determine and maintain).

Minor discrepancies in the presented distributions between the total driving time at the set speed and the total length of the measurement sections have no significant impact on the quality of the collected test data. Their primary source is omitting the bottom range of driving speed (e.g. speed below 2,5 km/h in Fig. 1 and 2) and exceeding the maximum speed by half the width of the speed range in the test (e.g. in Table 2, $V_{max} = 82,5$ km/h is more significant than $V_{rep} = 80$ km/h).

Conducting testing in line with the set distributions reduces the influence of the test driver's driving style on the measured load values, which increases the reliability of the data obtained. Moreover, it makes it possible to check whether the tested vehicle meets the assumed requirements concerning the movement speed on the identified surfaces of the test roads in relation to the assumptions included in the concept of operational use of the vehicle, while at the same time meeting the limits resulting from the binding regulations concerning permissible doses of vibrations and harmful factors in the driver's working environment [23]. It also gives a better reflection of the range of loads acting on the established components of the vehicle.

Furthermore, the average and maximum speed of the tested vehicle can be determined from the obtained vibration spectrum, also at a distance shorter than the one assumed in the standard [13, 16], by scaling the distributions in such a way that the total time of the driver's exposure to vibrations complies with the current regulation [18].

References

1. Allied Environmental Conditions and Test Publication – AECTP-240–1, NATO Standardization Agency (2009)
2. International Standard ISO 2631: Mechanical vibration and shock-evaluation of human exposure to whole body vibration. Part 1, General requirements. Geneva, Switzerland: International Organization for Standardization. ISO 2631-1 (1997)
3. Jamroziak, K., Kosobudzki, M., Ptak, J.: Assessment of the comfort of passenger transport in special purpose vehicle. Eksploatacja i Niezawodnosc - Maintenance Reliab. 15(1), 25–30 (2013)
4. Catalogue of standards for the operation of land-based technology – DTU-4.22.13.1(A), Bydgoszcz (2019) (in Polish)
5. Kosobudzki, M., Stańco, M.: Problems in assessing the durability of a selected vehicle component based on the accelerated proving ground test. Eksploatacja i Niezawodnosc – Maintenance Reliab. 21(4), 592–598 (2019). http://dx.doi.org/10.17531/ein.2019.4.8
6. Kosobudzki, M.,Stanco, M.: The experimental identification of torsional angle on a load-carrying truck frame during static and dynamic tests. Eksploatacja i Niezawodnosc – Maintenance Reliab. 13(9), 285–290 (2016). http://dx.doi.org/10.17531/ein.2016.2.17
7. Kosobudzki, M., Smolnicki, T.: Generalized vehicle durability index for different traffic conditions. In: AIP Conference Proceedings, vol. 2078, pp. 1–6 (2019)
8. Lee, Y.-L., Pan, J., Hathaway, R., Barkey, M.: Fatigue Testing and Analysis Theory and Practice. Elsevier Butterworth-Heinemann, Burlington (2005)
9. Standard BN-79/3615-01: Testing of motor vehicles and trailers. General guidelines (in Polish)
10. Defense Standard NO-23-A003/A1:2013 - Military Vehicles. Terminology (in Polish)
11. Standard PN-91/S-04100 - Vibrations. Methods of testing and evaluation of mechanical vibrations at workstations in vehicles (in Polish)
12. Standard PN-ISO 8855:2018-4 - Road vehicles. Dynamics and behavior during driving. Terminology (in Polish)
13. Defense Standards NO-06-A101-A108 - Armaments and Military Equipment. General technical requirements, inspection and testing method (in Polish)
14. Announcement of the Minister of National Defense of 24 July 2015 on the list of certification bodies and research units which have been granted accreditation in the field of defense and security together with the scope of accreditation, Official Journal of the Ministry of Defense of 29 July 2015, item 230 (in Polish)
15. Płaczek, M, Piszczek, Ł.: Testing of an industrial robot's accuracy and repeatability in off and online environment. Eksploatacja i Niezawodnosc – Maintenance Reliab. 20(3), 455–464 (2018). http://dx.doi.org/10.17531/ein.2018.3.15
16. Defense Standardization Manual. Military vehicles. Reliability tests PDNO-23-A503:2017 (in Polish)
17. Speed of vehicles in Poland in 2015. Session I. National Road Safety Council, Warszawa (2015) (in Polish)
18. Regulation of the Minister of Economy and Labour on safety and hygiene at work related to exposure to noise or mechanical vibration, Journal of Laws No. 157, item 1317 and 1318 of 05.08.2015 (in Polish)

19. Simiński, P.: Security of military vehicles. Sci. J. WSOWL **159**(1), 224–238 (2011)
20. The structure of road transport of the Polish Armed Forces. General Staff. Warszawa (1999)
21. Act on "Road Traffic Law", Journal of Laws No. 98, item 602 of 20.06.1997 (in Polish)
22. Preliminary Tactical and Technical Assumptions for a vehicle of medium payload of high mobility. SZRP Support Inspectorate, Warsaw (2011) (unpublished)(in Polish)
23. Zając, P.: Evaluation of automatic identification systems according to ISO 50001. In: Progress in Automation, Robotics and Measuring Techniques: Control and Automation, Advances in Intelligent Systems and Computing, vol. 350, pp. 345–355 (2011)
24. Rules and Organization of Operation and Repair of Equipment in Field Conditions - Instruction DD/4.22.10 - Inspectorate of the Armed Forces Support, Ministry of National Defense), Bydgoszcz (2013)(in Polish)
25. Zieliński, R.: Applied mathematical statistics. Elements. Centre for Advanced Studies at Warsaw University of Technology, Warszawa (2011)(in Polish)

Identification of the Mathematical Model of the Bucket Conveyor

Gabriel Kost[1]([⊠]) [iD], Sebastian Jendrysik[2] [iD], and Agnieszka Sękala[1] [iD]

[1] Faculty of Mechanical Engineering, Department of Engineering Processes Automation and Integrated Manufacturing Systems, Silesian University of Technology, Konarskiego 18A, 44-100 Gliwice, Poland
Gabriel.Kost@polsl.pl
[2] KOMAG Institute of Mining Technology, Pszczyńska 37, 44-100 Gliwice, Poland

Abstract. The article discusses the issue of identification of the control object, which is a bucket conveyor used to transport loose material of the heterogeneous gravimetric composition. The purpose of the identification process is to define a mathematical model of the actual bucket conveyor operating in the hard coal enrichment plant to optimize the energy consumption of its supply system. Observation of the actual working conditions of bucket conveyors shows that the basic parameter that affects the uniformity of the conveyor's operation, and thus influences its mechanical durability and its energy consumption optimality during the feeding phase is the speed of a bucket movement which is unevenly loaded with the transported stone material. The identification was carried out on a research stand using a real bucket conveyor working in a hard coal enrichment plant.

Keywords: Bucket conveyor · Mathematical model · Energy consumption · Control system

1 Introduction

Bucket conveyors are used in many industries to transport dry, often dusty materials under large elevation angles. One of the characteristic features of the bucket conveyor operation is the variability of its load, resulting from the variability of gravimetric composition of the material. In the event of an increase in the amount of material fed onto the conveyor, the speed should always be increased immediately in order to adapt to new loading conditions and to prevent excessive accumulation of material in its loading zone [2–4]. The identification task consists in such a modelling of the real object or control process under investigation that its structure and parameters can be determined on the basis of input and output signals obtained as a result of the identification experiment. The aim of this process is to find a mathematical model of a real control object that best describes the relationship between its input (forcing) and output values. The presented example of the identification process concerns a bucket conveyor (Fig. 1, [1]).

The identification task was carried out on a test stand using a real bucket conveyor operating in the hard coal enrichment plant. For this purpose, a special measuring system has been designed to measure the energy consumption of the conveyor and its

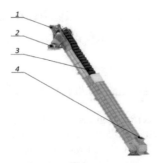

Fig. 1. Diagram of the bucket conveyor: 1 - material dumping zone, 2 - drive, 3 - chain with buckets, 4 - material loading zone [1].

temporary operating parameters. The data collected was used in models: ARX (Auto-Regresive eXogenous), ARMAX (Auto-Regresive Moving Average eXogenous), OE (Output-Error) and BJ (Box-Jenkins) [5–11, 13, 14]. As a criterion for model matching, the mean square error (NRMSE) was adopted, which allowed to determine the mathematical model with the best matching to the examined bucket conveyor with adjustable speed of bucket movement as a function of their bulk material load [1, 2, 5]

2   Assumptions

The identification process presented here has been carried out under the assumption that the conveyor is fed in the lower loading zone (Fig. 1, position 4) depending on the degree of tilt of the gate valve in the bulk container. The intensity of the material supply to the buckets depends on:

- the degree of opening of the hopper discharge slot - u_1 (Fig. 1), under which the buckets move,
- the speed of movement of buckets under the chute - u_2 (Fig. 1), which affects the power developed by the actuator drive.

Additionally, it was assumed that:

- It is not possible to fully and accurately determine the load of buckets with bulk material (its distribution, temporary changes in humidity),
- The resistance to bucket movement results from the bucket load and the degree of wear of the drive elements,
- Momentary disturbances during the conveyor operation process (model identification) are not known.

It is proposed that the planned identification process should be based on a MISO-type process model (Fig. 2).

Fig. 2. MISO model of the tested system.

where:

u_1 - opening of the jig release slot [%],
u_2 - linear speed of the moving buckets of the conveyor [*m/s*],
y_1 - mass of the material on the conveyor [*kg*].

It has also been assumed that the speed of bucket movement is a critical parameter in the process of conveyor control and should depend on the momentary load of the conveyor (buckets) with a stone material of heterogeneous structure.

3 Structure of the Measuring System

In the identification process a measuring system was used, the structure of which is shown in Fig. 3. A BTL5 magnetostrictive sensor (so called waveguide) was used to measure the degree of opening of the discharge gap of the bulk hopper, and an inductive sensor installed on the bucket driving wheel was used to measure the linear velocity u2 of the conveyor. The data in the measuring system was transmitted via Ethernet.

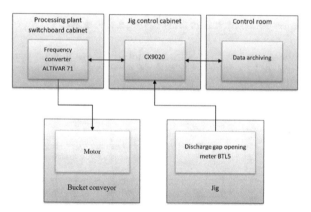

Fig. 3. Structure of the measuring system [1], where: ALTIVAR 71 – Schneider Electric frequency converter, CX9020 – Beckhoff driver.

The sampling frequency of the test object during measurements was determined according to the relation:

$$f_p = \frac{1}{\tau_p} \geq 2f_{max}[Hz] \tag{1}$$

where:

f_p - sampling frequency [Hz],
τp - sampling period [s],
f_{max} - maximum frequency occurring in the sample signal spectrum [Hz].

4   Model Identification

Because of the simplicity, it is assumed that a linear model of the conveyor is desired:

$$y_1(i) = \sum_{n=1}^{2} G_n\left(z^{-1}\right)u_n(i) + v(i) \tag{2}$$

where:

$G_n(z^{-1})$ - control path transmittance,
$u_n(i)$ - discrete sequence of input signal values,
$y_1(i)$- discrete sequence of output signal values,
$v(i)$ - discrete sequence of values constituting a total disturbance brought to the object's output.

Disturbances in the identification process are described by a random function:

$$v(i) = H\left(z^{-1}\right)e_{bs}(i) \tag{3}$$

where:

$H(z^{-1})$ - transmittance of the interference path,
$e_{bs}(i)$ - a discrete sequence of values constituting white noise.

Thus, the originally established MISO model (Fig. 2) of the conveyor takes the form (Fig. 4):

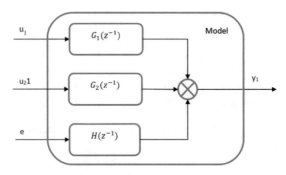

Fig. 4. MISO conveyor model.

Four standard models were adopted for the identification analysis [1, 2, 7, 8, 12, 13, 15–17]:

- **ARX** (Auto-Regresive eXogenous):

$$y_1(i) = \sum\nolimits_{n=1}^{2} z^{-k} \frac{B(z^{-1})}{A(z^{-1})} u_n(i) + \frac{1}{A(z^{-1})} e(i) \tag{4}$$

- **ARMAX** (Auto-Regresive Moving Average eXogenous):

$$y_1(i) = \sum\nolimits_{n=1}^{2} z^{-k} \frac{B(z^{-1})}{A(z^{-1})} u_n(i) + \frac{C(z^{-1})}{A(z^{-1})} e(i) \tag{5}$$

- **OE** (Output-Error):

$$y_1(i) = \sum\nolimits_{n=1}^{2} z^{-k} \frac{B(z^{-1})}{F(z^{-1})} u_n(i) + e(i) \tag{6}$$

- **BJ** (Box-Jenkins):

$$y_1(i) = \sum\nolimits_{n=1}^{2} z^{-k} \frac{B(z^{-1})}{F(z^{-1})} u_n(i) + \frac{C(z^{-1})}{D(z^{-1})} e(i) \tag{7}$$

where:

$y_1(i)$ – a discrete sequence of output signal values - in short, an output string or output,
$u_n(i)$ – a discrete sequence of input signal values - in short, an input string, an input or a excitation,
$e(i)$ – a discrete sequence of white noise values that disturb the object,
z^{-k} – the delay (backward displacement) of a signal by value of k so that $z^{-k} u_n(i) = u_n(i-n_k)$; n_k is called the (discrete) delay time and takes integer values greater than or equal to 1,

$z^{-k} \frac{B(z^{-1})}{A(z^{-1})} u_n(i)$, $z^{-k} \frac{B(z^{-1})}{F(z^{-1})} u_n(i)$ – control channels, $\frac{1}{A(z^{-1})}$, $\frac{C(z^{-1})}{A(z^{-1})}$, $\frac{C(z^{-1})}{D(z^{-1})}$, – interference channels modeling unmeasurable stochastic interference, operating in the object in the form of white noise, filtered by appropriate transmittance,
$A(z^{-1})$, $B(z^{-1})$, $C(z^{-1})$, $D(z^{-1})$, $F(z^{-1})$ – polynomials of the model, where:

$$A(z^{-1}) = 1 + a_1 z^{-1} + a_2 z^{-2} + \ldots + a_n z^{-n_n}$$

$$B(z^{-1}) = 1 + b z^{-1} + b_2 z^{-2} + \ldots + b_n z^{-n_b+1}$$

$$C\left(z^{-1}\right) = 1 + c_1 z^{-1} + c_2 z^{-2} + \ldots + c_n z^{-n_c}$$

$$D\left(z^{-1}\right) = 1 + d_1 z^{-1} + dz^{-2} + \ldots + d_n z^{-n_d}$$

$$F\left(z^{-1}\right) = 1 + f_1 z^{-1} + f_2 z^{-2} + \ldots + f_n z^{-n_f}$$

Examples of the results of the identification process are shown in Fig. 5, 6, 7, 8, 9, 10, 11, 12, 13, 14, 15 and 16.

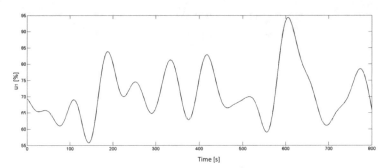

Fig. 5. Opening degree of discharge slots.

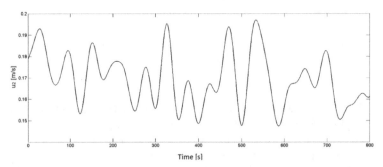

Fig. 6. Bucket conveyor speed graph.

A) **The waveforms of input and output variables used model identification**
B) **Verification analysis of the ARX model**
C) **Verification analysis of the OE model**
D) **Verification analysis of the BJ (Box-Jenkins) model**

4.1 Assessment of the Model Structure

At the stage of model structure selection, degrees of polynomials were determined for polynomials were determined for all elements from the set of models, created from

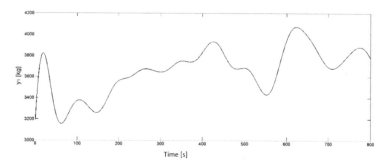

Fig. 7. Weight of material on the conveyor.

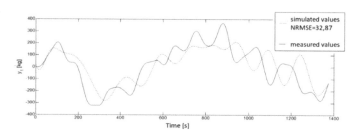

Fig. 8. Comparison of measured and simulated values.

Fig. 9. Autocorrelation of the residuals.

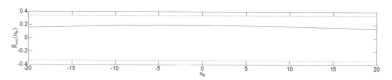

Fig. 10. Cross-relation of the residuals.

possible combinations of degrees of differential polynomials n_a, n_b, n_c, n_d, n_e, n_f taking into account the assumption that the degree of differential polynomials of the model does not exceed 4. This assumption results from the need to implement the algorithm in the PLC.

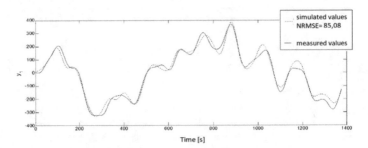

Fig. 11. Comparison of measured and simulated values.

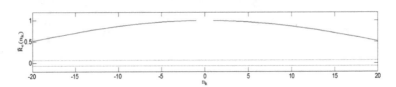

Fig. 12. Autocorrelation of the residuals.

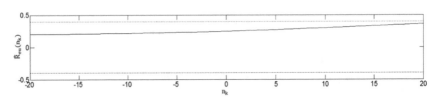

Fig. 13. Cross-relation of the residuals.

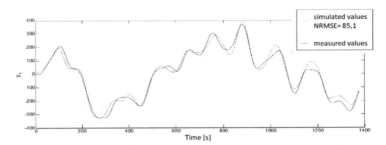

Fig. 14. Comparison of measured and simulated values.

The quality of the results obtained was evaluated [1, 15–17]:

- Akaike AIC information criterion:

$$AIC = -2log\left(\hat{L}\right) + 2p \qquad (8)$$

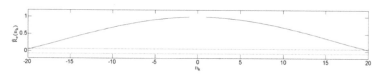

Fig. 15. Autocorrelation of the residuals.

Fig. 16. Cross-relation of the residuals.

- Bayes BIC information criterion:

$$BIC = -2log\left(\hat{L}\right) + p \cdot log(N) \tag{9}$$

- FPE prediction error criterion:

$$FPE = \frac{N+p}{N-p}V_{pe} \tag{10}$$

- FOE output error criterion:

$$FOE = \frac{N+p}{N-p}V_{oe} \tag{11}$$

where:

p – number of model parameters,
N – number of signal samples,
$\hat{L}$ – the estimated probability of obtaining such an observation value, what was really obtained,
V_{pe} – the loss function as defined by the equation $\frac{1}{N}\sum_{i=1}^{N}(y_i - \bar{y}_i)(y_i - \bar{y}_i)^T$,
V_{oe} – the loss function as defined by the equation $\frac{1}{N}\sum_{i=1}^{N}(y_i - \bar{y}_i)^2$.

The most representative models of each type were then analysed in terms of their ability to predict the actual process. The matching of the models to the actual object was based on the NRMSE normalized root mean square error:

$$NRMSE = \left[1 - \frac{\|y - \hat{y}\|}{\|y - \bar{\hat{y}}\|} \right] \cdot 100\% \qquad (12)$$

where:

y_1 – a discrete sequence of the output variable value,
$\hat{y}$ – a discrete sequence of the output variable value from the model,
$\bar{y}$ – a discrete sequence of the average output variable value.

A residual analysis was also carried out. The residual series is formed as the difference between the measured variable values and the forecast values. The mean value of the series should be close to zero, although at the same time its standard deviation need not be zero. For the purpose of the residual analysis, the autocorrelation and cross-correlation of the model error were plotted for the designated models *(i)*.

$$\varepsilon(i) = y_1(i) - \hat{y}(i) \qquad (13)$$

Autocorrelation is described by the equation:

$$\hat{R}_e(n_k) = \frac{1}{n} \sum_{i=1}^{n} \varepsilon(i)\varepsilon(i - n_k) \qquad (14)$$

Cross-relation defines the correlation between the model error and its input, it is defined by the equation:

$$\hat{R}_{eu}(n_k) = E\varepsilon(i)u_{12}(i - n_k) \qquad (15)$$

where:

$E(x)$ – expected value x,
n_k – discreet delay time,
u_{12} – input variable,
(i) – model error.

The results of these analyses are shown in Fig. 17, 18, 19 and 20. The analysis was carried out on the basis of the graphs, in which the 95% confidence intervals were marked. The model can be considered satisfactory when all residuals are in the confidence range.

An example comparison of specific qualitative indicators for the ARX models and the adopted degrees of polynomials (na, nb, nc, nd, ne, nf) is shown in Table 1. The same was carried out for the other models: ARMAX, OE and BJ.

Using the concept of residual analysis, the autocorrelation and cross-correlation of the model error were plotted.

From the group of analyzed cases, the one with the lowest value of the *FOE* criterion was always selected. The mathematical description of the model presents the equation:

$$y_1(i) = \frac{-0.00001103}{1 - 2.968z^{-1} + 2.938z^{-2} + 0.9701z^{-3}} u_{12}(i) \qquad (16)$$

The final results of the analysis are shown in Fig. 17.

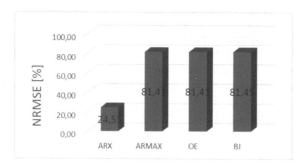

Fig. 17. Average model matches determined for 3 data groups.

Table 1. Comparison of structure selection criteria determined for different degrees of polynomials of the ARX model.

ARX model						
No.	Structure		AIC	BIC	FPE	FOE
	n_a	n_b				
1	1	1	9.5526	15.7733	14081.4565	713.4603
2	1	2	7.9821	27.9127	2928.1699	61.2731
3	1	3	7.6579	31.3376	2117.3538	410.6816
4	1	4	7.6546	38.3576	2110.3781	456.1131
5	2	1	9.7542	12.6891	17226.5220	123888.8326
6	2	2	10.4337	20.8821	33985.1209	76540.3225
7	2	3	8.1671	42.0935	3523.1712	70.4414
8	2	4	8.1404	46.9852	3430.2061	226.7515
9	**3**	**1**	**10.5314**	**36.9115**	**37472.2786**	**25.3020**
10	3	2	9.4008	34.3636	12097.6965	3360.1062
11	3	3	7.2325	38.7957	1383.6182	13610.6757
12	3	4	7.8572	49.7548	2584.1562	2109.0252
13	4	1	13.5381	35.0963	757750.0664	2329.5038
14	4	2	9.7809	40.1782	17693.3515	6818.3855
15	4	3	8.4385	46.1469	4621.3768	12809.6926
16	4	4	6.8817	59.0531	974.2676	749.7801

5 Conclusions

Identifying the actual object is a difficult and complex process. Its effectiveness depends on proper physical identification of the actual device. Therefore, a possible way of obtaining a mathematical model of an object may be its dynamic model. However, the constructional complexity of real devices most often makes it impossible to precisely define all the dynamic relationships taking place in an object. Therefore, the use of recognized, general linear object models proposed by automation is often the best way to effectively obtain a fairly accurate model of the analyzed control object. The adopted qualitative criteria may be helpful in this process to assess the quality of the representation of the real system into a mathematical model.

In the process carried out and described in the work it was established that:

- the best average match, determined with reference to the data recorded during the identification experiment, was obtained for the BJ model,
- the autocorrelation graphs of the prediction residuals indicate that the process dynamics are not sufficiently reproduced, it may be related to insufficient filtering of variable y_l from high frequencies, which is connected with the way it is measured and the operation of the conveyor gear,
- all models have residuals within the confidence limits, and the residuals of the BJ model are the smallest. It was therefore concluded that of all analysed models, BJ best reproduces variable y_l.

References

1. Jendrysik S., Kost G.: Method of controlling the bucket conveyer in a jig beneficiation node. Phd Monograph, KOMAG, (Gliwice) (2018) [in polish]
2. Jendrysik, S., Kost, G.: Control of bucket Conveyor's output. In: Świder, J., Kciuk, S., Trojnacki, M. (eds.) Mechatronics 2017 - Ideas for Industrial Applications. MECHATRONICS 2017. Advances in Intelligent Systems and Computing, vol 934. Springer, Cham (2019
3. McBride, W., Sinnott, M., Cleary, P.W.: Discrete element modelling of a bucket elevator head pulley transition zone. Granular Matter **13**, 169–174 (2011). https://doi.org/10.1007/s10035-010-0243-2
4. Ma, F., Xiong, F., Li, G.: Key structure innovation and optimization design of bucket elevator. In: Wang, K., Wang, Y., Strandhagen, J., Yu, T. (eds.) Advanced Manufacturing and Automation VIII. IWAMA 2018. Lecture Notes in Electrical Engineering, vol 484. Springer, Singapore (2019
5. Seborg, D.E., Mellichamp, D.A., Edgar, T.F., Doyle III, F. J.: Process Dynamics and Control. John Wiley & Sons, Hoboken (2010)
6. Starr, G.: System identification. Introduction to Applied Digital Controls. LNEE, pp. 177–187. Springer, Cham (2020). https://doi.org/10.1007/978-3-030-42810-5_8
7. Morari, M., Lee, J.H.: Model predictive control: past, present and feature. Comput. Chem. Eng. **23**, 667–682 (1999)
8. Maciejowski, J.M.: Predictive Control with Constraints. Prentice Hall, London (2002)
9. Box, G.E.P., Jenkins, G.M.: Time Series Analysis: Forecasting and control, Rev edn. Holden-Day, San Francisco (1976)
10. Ljung, L.: State of the art in linear system identification: time and frequency domain methods. In: Proceedings of the American Control Conference, vol. 1, pp. 650–661 (2004)
11. Muhammad, Z., Yusoff, Z.M., Rahiman, M.H.F., Taib, M.N.: Modeling of Steam Distillation Pot with ARX Model. In: IEEE 8th International Colloquium on Signal Processing and its Application (2012)
12. Ljung L.: System Identification Toolbox For Use with MATLAB. MathWorks (2004)
13. Rosenqvist, F., Karlstrom, A.: Realisation and estimation of ¨ piecewise-linear output-error models. Automatica **41**(3), 545–551 (2005)

14. Breschi, V., Bemporad, A., Piga, D., Boyd, S.: Prediction error methods in learning jump armax models. In: IEEE Conference on Decision and Control (CDC), pp. 2247–2252 (2018)
15. Królikowski A., Horla D.: Identification of control objects, discrete parametric methods. The Publishing House of Poznan University of Technolog, Poznań (2010) [in polish]
16. Kasprzyk J.: Process identification. Publishing House of the Silesian University of Technology, Gliwice (2002) [in polish]
17. Osowski S.: Modeling and simulation of dynamic systems and processes. Publishing House of the Warsaw University of Technology, (Warsaw) (2007) [in polish]

Selected Problems of the Numerical Capacity Assessment for Floating Systems

Wiesław Krasoń[(⊠)] [iD]

Institute of Mechanics and Computational Engineering, Military University of Technology,
Warsaw, Poland
wieslaw.krason@wat.edu.pl

Abstract. Modular floating systems are structures used both for the army and the civilian population needs. They enable construction of temporary water crossings, for example, in place of permanent out of service bridges or objects renovated due to mechanical and other damages. Special features of this type structures include the possibility of using them regardless of the depth and terrain conditions of the water crossings. The requirements currently set for such structures force their modernization as well as search for innovative solutions. They should allow easy and quick assembly-disassembly of the structure, and have a low weight and dimensions to facilitate transport as well as minimize storage space. These requirements are met by a pontoon construction based on an innovative cassette designed in Poland (Military University of Technology). A single cassette contains a high-strength elastic coating pontoon which, after filling with air, allows obtaining the regulation of buoyancy force, appropriate depending on the type of water crossings. The paper deals with testing the load capacity of a pontoon bridge composed of repetitive segments-cassettes with adjustable immersion. Numerical models of a single floating cassette and a group of connected cassettes constituting ferries or sections of a pontoon bridge are also presented in the paper. The methodology for load capacity tests and selected test results of such floating objects are discussed. Multibody numerical methods and FEM (Finite Element Method) are used for dynamic and static analysis of multi-component floating systems. Selected field tests and laboratory stand tests conducted with the use of the strength machine are applied to validate the various type of numerical models.

Keywords: Prototype modular multi-segment system · Research methodology · Numerical and experimental tests · Multibody-rigid solid models · FE models

1 Introduction

The occurrence of clearances and the effects of their interactions in even precisely designed and manufactured mechanisms are in fact difficult to avoid. It is therefore necessary to work on eliminating or minimizing undesirable aspects of the clearance occurrence. For this purpose, it is required to explain and understand phenomena related to the impact of clearances. Experimental tests of complex structures with clearance are difficult and expensive. Theoretical methods have some limitations, due to geometrical

A. Mężyk et al. (Eds.): SMWM 2020, AISC 1336, pp. 168–180, 2021.
https://doi.org/10.1007/978-3-030-68455-6_15

nonlinearities determined by clearances. Therefore, simulation methods for tests using numerical analyses of multipart systems with clearances are gradually improved.

The basic method for assessing technical facilities is to analyze their strength. Such facilities include multi-section constructions, modular structures and a wide range of machine components and complete devices for various purposes, in which the proper cooperation of individual components plays a significant role. A typical approach to complex structures (i.e. mechanical systems) is the analysis of selected parts of the structure. Correct assessment of the entire structure effort requires the use of an appropriate research methodology, including a correct determination of boundary conditions.

The correct cooperation of elements of complex structure systems requires selection of a proper matching of cooperating parts and taking into account an appropriate value of clearance. Determining the desired clearance may be a challenging task related to optimization studies, multi-criteria analyses and a need to use tedious multi-variant research. It is connected with a long research time, high costs and a necessity to apply advanced methods and specialized tools, e.g. programming, to such works.

Multivariate strength tests that would show an influence of the presence of clearances and their size on the effort of the structure are often inaccessible and difficult to perform (especially experimental research), due to high costs or ensuring the safety of the device itself, its equipment and the servicing team. All the above remarks are particularly important in the case of large-size structures. Such objects include bridges (floating, folding, and others) for both military and civilian use, and also special-purpose, costly in design and construction, structures on which the safety and people's lives often depend. This applies to, among others, aircraft landing gears, suspensions of special vehicles or other original solutions in the transport field.

The analysis of the clearance impact on the strength of engineering structures is the purpose of this study. The basic element of the presented scientific investigations is the numerical capacity assessment of the multi-segment floating system. The methodology of modeling as well as the selected aspects of numerical analysis of objects with constructional clearances are discussed in detail on the example of this prototype solution [1–3].

2 Subject of Research and Methodology

The subject of the research presented in this work is a multi-layer prototype floating system composed of cassettes (Fig. 1) with the adjustable buoyancy. In the work, a multi-layer floating system is meant as a set of the same floating modules with a system of joints and additional equipment enabling construction of floating objects in various construction configurations. The system of side as well as bow and stern connections of a single module allows assembling the floating bridges in various configurations, such as a single, mixed and double ribbon, as well as building ferries, floating platforms, coast bridges or floating quays with different length and width.

Modular floating systems are structures used both for the army and the civilian needs. They enable construction of temporary water crossings, for example, in the place where permanent bridges are out of service or objects are renovated due to mechanical and other damages. Bridges and floating ferries are used for the evacuation of people and

a)

b)

c)

Fig. 1. Prototype floating system composed of cassettes: a) closed single cassette, b) opened single cassette, c) a set of two joined cassettes at a field stand for symmetric bending tests on fixed supports (1 ÷ 8 – the load number used in sequence during the field test).

for the removal of the effects of natural disasters as well. Special features of this type structures include a possibility to use them regardless of the depth and terrain conditions of the water crossings.

The requirements currently set for such structures force their modernization as well as search for innovative solutions. They should allow an easy and quick assembly-disassembly of the structure as well as should have a low weight and dimensions to facilitate transport and to minimize the storage space. These requirements are met by a pontoon structure based on an innovative cassette designed at the Military University of Technology. A single cassette contains a high-strength elastic coating-pontoon which, after filling with air, allows obtaining an appropriate buoyancy force depending on the type of water crossings.

The numerical assessment of the bridge' strength, even in simplified models (2D rigid solid models and 2D FE models), creates the basis for a low cost and relatively quick diagnosis of the performance parameters of such a multi-segment system or its safety testing, based on computer simulations and a detailed analysis of parameters describing an effort.

To analyze the strength of floating bridges, mainly theoretical methods (in the country [4–6]) or simplified models and specialized software, taking into account the replacement geometrics and physical characteristics of such floating objects, are used.

The methodology of the continuous description application in the static and dynamic calculations of floating bridges was developed by the authors of publications [6, 7] 1980s. The results of theoretical analyses, developed with their use, found their industrial application. However, this approach is not devoid of certain shortcomings, including a difficulty describing precisely the phenomena at the interface between the short circuit and the opening zones of the bridge. The methodology can be successfully applied to the analysis of objects assembled from identical pontoons. In the case of bridges with a stepwise variable stiffness along the longitudinal axis of the structure, its direct use is impossible. This applies to mixed floating ribbons commonly used in the army [8, 9], convenient for use in changing tactical and economic conditions (flexible and inexpensive to match the type of equipment). Discrete approach methods are suitable for testing bridges in mixed configurations, with the representation of different clearances in individual joints of such multi-segment systems. Discrete methods are susceptible to algorithmization and easy to implement.

The numerical methods based on the finite element method and simple two-dimensional floating bridge models were initiated at the Faculty of Mechanical Engineering of the Military University of Technology (MUT) [10, 11]. This work is a continuation of the research into the development, improvement and implementation of original and modern designs of floating bridges [2, 12] in our country.

3 Research Methodology

Experimental analysis of the strength of a floating bridge is possible; however, it is used only to a limited extent due to its high costs and time consumption. The theoretical analysis of the bridge effort using numerical methods is therefore required as an important stage in the assessment of the safety state of the crossing both in typical operating conditions as well as in unusual conditions, particularly to ensure continuous operation of these structures exposed to frequent damage in combat conditions.

Taking into consideration the above mentioned facts, the methodology for testing the strength of a novel floating bridge with adjustable displacement in a numerical and experimental approach a prototype research object with a patented system of side connections is an original scientific and engineering problem. The assessment of the bearing capacity of such floating objects in various configurations is also an indispensable stage of the implementation work of the floating system being developed at MUT.

The methodology was developed for testing the strength of not only special bridges, but also other, more widely understood multipart, systems combined taking into consideration the interaction of clearance in joints. It enables numerical and experimental tests of the strength of such newly designed systems at the stage of their preliminary tests and implementation works. The methodology also includes tests on the strength of the multi-part systems in service/operation, structures under modernization or showing signs of wear (for example, connected with an increase in clearances in joints between the components/elements) as well as objects qualified for renovations. The proposed methodology is used and tested in [1] on the example of strength and capacity determination of the prototype cassette structure of a floating bridge with the adjustable displacement [12].

The methodology for strength analysis of multi-segment systems proposed in the work is presented in the form of a block diagram in Fig. 2. The procedure for testing the bearing capacity of a floating single-ribbon bridge, assembled from prototype cassettes with the variable buoyancy, complements this methodology. It is discussed in the form of a diagram in Fig. 3.

The methodology for testing multi-segment systems with clearances proposed in [1] enables the analysis of the strength of complex structures and mechanisms, including multiple moving connections and contact phenomena.

4   Selected Aspects of Numerical Analysis of the Floating Systems Strength

The proposed methodology takes into account different experimental and numerical approaches in the research of floating bridges. The use of experimental methods in the test of floating systems is cost consuming. They may be often used only to a limited extent for security reasons. Experimental research may concern complete floating systems, including, for example, floating bridge ribbons, a ferry, etc. Research of such a scale is expensive and requires an appropriate infrastructure (e.g. a training ground with a water container of an appropriate size) as well as an extensive research facilities. Most frequently, this type of field tests [1, 12–16] are organized in the form of qualifying or receiving tests of a ready-to-use floating ribbon bridge. Descriptions of the strength tests of structural fragments (e.g. a single pontoon [17], a set of two prototype cassettes [15]) or stand tests of subassemblies, parts separated from the entire structure [1, 18] and laboratory material tests [1, 18] are available.

Experimental studies of such structures may be also conducted in relevant laboratory conditions with the use of reduced physical models with maintaining an appropriate scale relative to real objects [19]. Experimental studies of this type are also subjected to significant limitations and may be considered only as additional supporting research.

Due to commonly known difficulties and limitations in the application of experimental methods in testing complex structures of large dimensions, including those loaded variably in time [1, 20–22], numerical methods and a computer technique for strength tests of floating systems [23] are used.

Numerical methods may be used even at the preliminary stage of the tests of such structures (Fig. 2). Due to the nature of the preliminary work, it is often necessary to perform verifying calculations repeatedly. In this case, analytical methods, which after proper adaptation (e.g. algorithmization) allow for the use of a computer technology for quick implementation of repetitive calculation sequences, are useful. As an example, an analytical method to assess the impact of clearances and to estimate kinematic parameters of a floating ribbon type with various configurations, proposed in [1], may be mentioned.

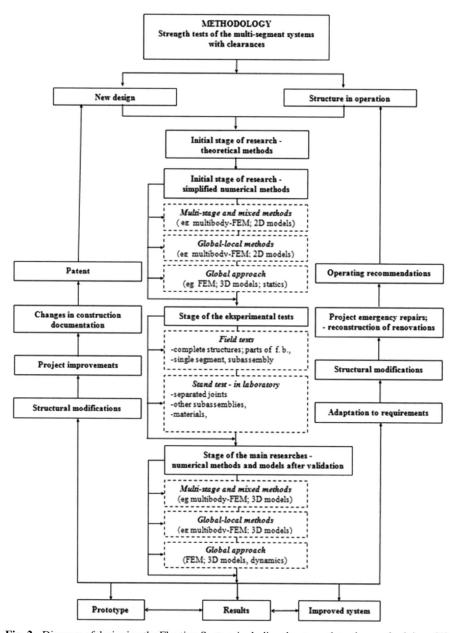

Fig. 2. Diagram of designing the Floating System including the strength testing methodology [1].

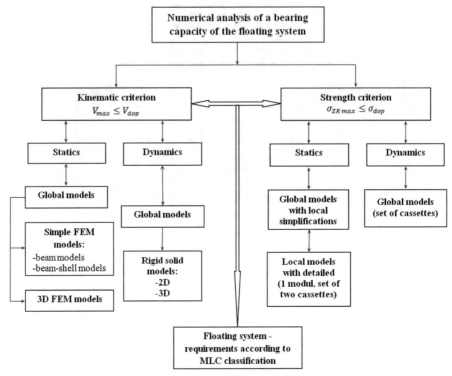

Fig. 3. Numerical research methodology for determining the bearing capacity of floating bridges [1].

Numerous load models, including time-varying models [20, 24], inertial mapping of moving vehicles (multibody rigid solid 3D model - Fig. 4) [1, 24], FE statics and impulse-type excitations (Fig. 5 and 6 – 3D FE models) are considered in multi-variant numerical analyses [20]. An influence of the size of clearances and various friction conditions in the rod joints on the strength of prototype cassettes and floating systems built from them is also determined [1].

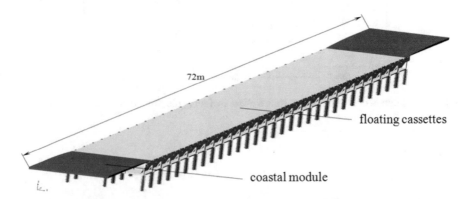

Fig. 4. Multibody numerical models (3D rigid solid) of the single ribbon type bridge.

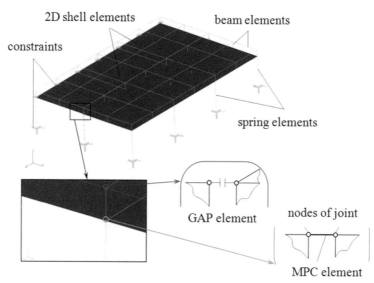

Fig. 5. 3D FE numerical models of the floating system of two cassettes with the constructional details explanation ('GAP element'– 'point to point' type contact element; 'MPC element' – multipoint constraint element for modelling of joints in the multi-segment structure; 'beam elements' – additional hypothetical model of bridge structure with reduced stiffness; 'spring elements' – reaction of water reduced in points of structure).

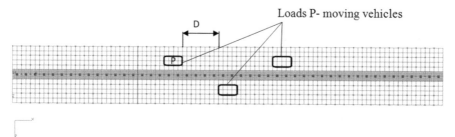

Fig. 6. 3D FE numerical models for statics and dynamics analyses with inertial mapping of moving vehicles (loads distance - D) and impulse-type excitations.

Among the numerical methods used in the initial stages of work on testing of the strength of floating systems with clearances, original own methods can be indicated (e.g. methods using 2D and 3D rigid solid models [1, 5]) using relatively simple quick mathematical models and those in which there are used popular methods (e.g. FEM) and proprietary or commercial software, adapted to simulate the operation of multiple structures, taking into account static and dynamic clearances [1, 12, 18] as well as simplified discrete models [1, 5, 11]. Such methods may also be based on multipart 2D rigid models [1] used in multibody analyses (Fig. 2).

176   W. Krasoń

In the part of the main research on floating systems, as indicated in the methodology (Fig. 2), mainly numerical simulations performed using 3D models, including commercial calculation codes based on FEM and multibody methods are taken into account [23, 25]. The various type of numerical models used at the stage of analyses were previously validated based on the results of the experimental studies [1, 18].

The results of the research (Fig. 7 and 8) may be used to build a prototype of a newly designed structure and to implement a preproduction batch or in the process of widely understood diagnostics and modernization of a structure in operation [1, 2, 12–15, 18].

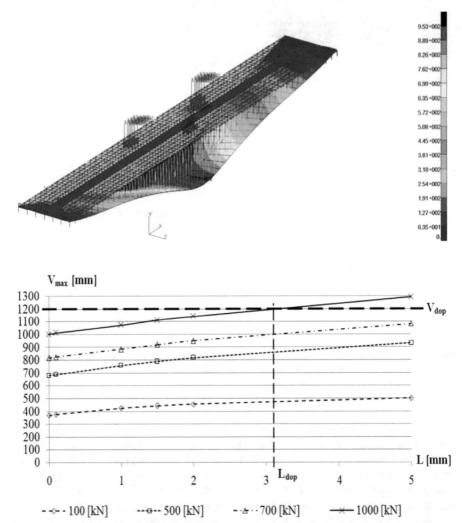

Fig. 7. Displacement map and deformation of the floating ribbon type bridge with length 96 m. Diagram with maximal displacements (Vmax) of ribbon bridge as a function of load (MLC – Military Load Classification or STANAG classification) and clearances (L and Ldop – the admissible-acceptable clearances) in joints of bridge - Vmax = f(L, P).

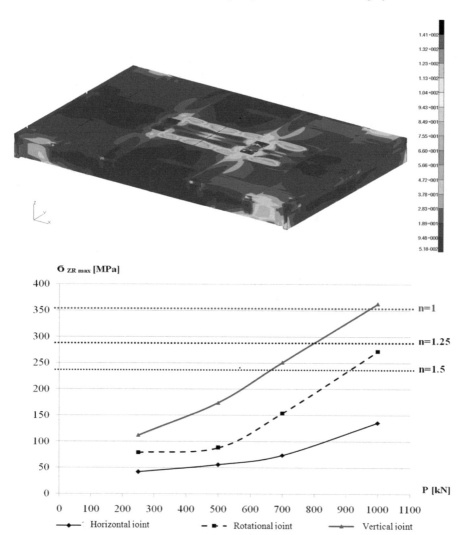

Fig. 8. Selected results of static analyses with determination of the impact of various loads (including standard ones defined according to STANAG) for the load capacity and strength of the prototype floating structures - maximal HMH stress (reduced with the Huber-Misses-Hencky strength hypothesis).

Performing numerical analyses of the impact of clearance and loads corresponding to the actual weights of the crossing vehicles on the bearing capacity of floating bridges in various operational configurations is an important stage in the study of the bridge prototype structure [1]. In such research, various methods, calculation programs and models may be used (Fig. 3). The methodology for the bearing capacity numerical tests of floating bridges provides for the use of various computer techniques, models with different levels of advancement expressed in the precision of mapping construction details and taking into account the broadest possible spectrum of external interactions. Based on the experiments and results of work related to implementation of this methodology, it was found that the choice of the methodology of analysis and numerical models should be closely matched to the assumptions and requirements adopted at a given stage of the bridge tests [1, 18]. It is assumed that in order to rationalize the proportion of work inputs to the development of models, the preparation of variants of the appropriate boundary and initial conditions and the duration of the analysis together with the subsequent process of processing the results, the strength test methodology for the multi-segment systems with clearances is shown in Fig. 2, and the detailed methodology for determining the limit resistance of such structures is described in Fig. 3. An important stage of the methodology is the use of the global 2D and 3D models of the simplified floating ribbons in the numerical studies [1].

FE models combining the features of the simplest 2D beam models and 3D models of floating ribbons with a high level of precision in the mapping of mechanical features are effective, especially in multi-variant studies (accurate mapping of individual modules, their stiffness and interaction conditions, including models of joints and clearances, etc.).

Among them are 3D FE models, in which a single floating module is replaced by a simplified shell-beam structure with a possibility of mapping the side connections and modeling the non-linear contact phenomena resulting from the occurrence of clearance in joints. 3D Models of this type enable FE analysis in terms of statics and dynamics, taking into account the spatial state of the load causing simultaneous bending in two planes and torsion the floating ribbon [1].

Global 2D, 3D rigid solid (named here multibody model) and 3D FE models (Figs. 2, 3 and 4, 5) are used in the tests of the impact of clearances on the strength of the floating ribbons built on the basis of prototype cassettes. Global 3D shell-beam FE models of ribbon fragments were tuned based on data from experimental research. After adjusting them and checking the correctness of operation in 3D models with a limited number of modules (fragments of a floating ribbon connected with two to ten individual cassettes), the models mapping typical ribbons, operated for example in combat conditions, with a length of about 100 m, in various construction configurations, were developed. Paper [1] discusses the problem of analysis of such ribbon bridges using simplified global 3D models, taking into account the effect of interaction in two planes (longitudinal bending and transverse torsion of the system), and results of static (Fig. 5, 6) and dynamic analyses with determination of the impact of various loads (including standard ones defined according to STANAG [26]) and clearances for the load capacity and strength of the prototype floating structures based on the cassettes jointed into various floating operating systems.

5 Conclusions

The research methodology allows for a diagnosis of the already implemented structures and verify their technical condition, correct operation in various phases of their service life, including planning the inspections, overhauls, analysis of the impact of immediate repair works on the structure or formulating the operational recommendations or verifying the adaptation and adaptation of the existing structure to implement new tasks and possibilities of meeting the reformulated technical requirements (e.g. in the case of military equipment for crossing operations) [1, 12–15].

In the numerical methodology of load capacity analysis of floating bridges, the simultaneous fulfillment of the kinematic and endurance criterion is verified in terms of statics and dynamics. The mixed numerical analysis methods are used for this purpose. Kinematic parameters, in this case immersion of floating objects, are determined in static and dynamic studies by means of global 2D and 3D models of complete bridge ribbons developed with different levels of mapping the construction details and mixed numerical methods. In order to analyze the effort and verify the strength criterion, 3D FE deformable models of a single cassette, a set of two cassettes and a separate spindle assembly of the side joint are used. These models are treated as local. They require the development of appropriate boundary and initial conditions for replacement; however, they enable the detailed mapping of structural solutions of the analyzed objects and a precise description of their operating conditions.

The advantages and originality of the methodology proposed in [1] are manifested in the fact that using simplified global 2D models of the floating system with clearances in joints and the local detailed 3D models of a separated joint it allows for analysis of the strength of floating modules with joints, taking into account the complex conditions of different loads, including variables in time, contact phenomena and variable conditions of cooperation of the connector elements.

It is possible to increase the efficiency (mainly in terms of numerical costs of the strength analysis of complex floating systems in comparison to numerical research using more precise 3D models for one-stage analyses of such objects. This was achieved by suitable combining of various research methods (experimental and numerical), particularly at the stage of tuning the selected characteristics and validation of the various type of numerical models, the use of multi-stage methods and appropriate models in the subsequent stages of numerical analysis.

Within the implementation of the described research methodology, numerical and experimental analyses of floating bridges with lengths corresponding to actual crossings are performed taking into account clearances and various load variants (including those defined according to STANAG military standards) and the load capacity of prototype floating structures in various construction systems. It was found that floating bridges built on the basis of prototype cassettes with maximum displacement (after fully filling the elastic tanks with air), in mixed and double ribbon systems, may operate safely in crossings of vehicles with a weight of 1000 kN (100 MLC according to NATO classification [26]) moving at a speed of about 5 mps. The allowable load capacity of a single prototype ribbon is 700 kN (70 MLC).

References

1. Krasoń, W.: Analiza wytrzymałości prototypowego wieloczłonowego systemu pływającego z luzami konstrukcyjnymi, Wydawnictwo WAT, pp. 280, ISBN 978-83-7938-237-8, Warszawa (2019)
2. Niezgoda, T., Krasoń, W., Derewońko, A., Chłus, K., Popławski, A.: Patent RP. PAT.223689. Kaseta mostu pływającego. Poland (2016)
3. Niezgoda, T., Krasoń, W., Derewońko, A., Chłus, K., Popławski, A.: European Patent. EP2570551. *A cassette of a floating bridge.* EU (2018)
4. Bursztynowski, Z., Wieczorek, M., Krasoń, W.: Porównanie wyników analizy statycznej mostów pływających otrzymanych w modelu ciągłym i dyskretnym. Biuletyn WAT, XLVII, 9, Warszawa (1998)

5. Krasoń, W., Wieczorek, M.: Analiza numeryczna mostu pływającego modelowanego elementami sztywnymi w systemie WSTĘGA i MSC/NASTRAN. Przegląd Mechaniczny, Nr 9, Warszawa (2001)

6. Beaufait, F.W., Hoadley, P.W.: Analysis of elastic beams on nonlinear foundations. Comp. Struct. 12, 669–676 (1980)

7. Langer, J.: Analiza dynamiczna przęsła mostowego obciążonego ruchomym pojazdem. Arch. Inż. Ląd. 20, 4 (1974)

8. Instrukcja Szefostwa Wojsk Inżynieryjnych: Park Pontonowy PP-64. Opis i użytkowanie. Wydawnictwo MON, Warszawa (1986)

9. MON: Mosty wojskowe. Szefostwo Wojsk Inżynieryjnych MON, Warszawa (1994)

10. Dacko, M., Borkowski, W., Dobrociński, S., Niezgoda, T., Wieczorek, M.: Metoda elementów skończonych w mechanice konstrukcji. Arkady, Warszawa (1994)

11. Wieczorek, M., Krasoń, W.: Numeryczna analiza mostów typu wstęgi z uwzględnieniem segmentu brzegowego, Biuletyn WAT, XLV, nr 7(527), pp. 107–121, Warszawa (1996)

12. Krasoń, W.: Koncepcja, rozwiązania konstrukcyjne i badania systemu pływającego o regulowanej wyporności. Inżynieria wojskowa - problemy i perspektywy, Wojskowy Instytut Techniki Inżynieryjnej, ISBN 978–83-948983-0-4, pp. 123–145. Wrocław (2018)

13. Krasoń, W.: Numeryczno-eksperymentalne badania mostów specjalnych, Inżynieria Wojskowa, problemy i perspektywy. Wojskowy Instytut Techniki Inżynieryjnej, Wydawnictwo Politechniki Wrocławskiej, ISBN978-83-911434-8-3, pp. 109–122. Wrocław (2014)

14. Krason, W., Malachowski, J.: Field test and numerical studies of the scissors-AVLB type bridge. Bull. Polish Acad. Sci. Techn. Sci. 62(1), 103–112 (2014)

15. Krason, W.: Strength analysis of the scissors-AVLB type bridge. Shell Structures: Theory and Applications, vol. 2, Taylor & Francis Group, pp. 307–310, London, UK (2010)

16. Bartnicki, A.: Badania wytrzymałościowe i trwałościowe prototypu mostu samochodowego MS-20. Biuletyn WAT, Vol. LIX, Nr 1, Warszawa (2010)

17. Kamyk, Z., Śliwiński, C.: Etapy rozwoju konstrukcji modelu aluminiowo-kompozytowego bloku pontonowego. Inżynieria wojskowa-problemy i perspektywy, monografia WITI, pp. 143–154, Wrocław (2014)

18. Krason, W., Popławski, A.: Numerical research of the cassette bridge joint strength with mapping of stand for experimental tests. In: AIP Conference Proceedings 2078, 020050 (2019). https://doi.org/10.1063/1.5092053, pp. 1–6, Published Online: 2019/03/04

19. Fu, S., Cui, W.: Dynamic responses of a ribbon floating bridge under moving loads. Marine Structures 29, 246–256 (2012)

20. Krason, W., Małachowski, J.: Multibody rigid models and 3D FE models in numerical analysis of transport aircraft main landing gear. Bull. Polish Acad. Sci. Techn. Sci. 63(3), 745–757 (2017)

21. Krasoń, W., Małachowski, J.: Dynamics analysis of the main landing gear in 3D model. J. KONES2006 Powertrain and Transport', vol. 14, no. 3, pp. 305–310 (2007)

22. Kowalewski, Z.L.: Kierunki i perspektywy rozwoju badań wytrzymałościowych. Wydawnictwo Instytutu Transportu Samochodowego, p. 227, Warszawa (2008)

23. Stankiewicz, M., Krasoń, W., Barnat, W.: Badania numeryczne odcinka mostu pływającego typu wstęga w modelach 3D. Miesięcznik Naukowo-Techniczny Mechanik Nr2 (2012)

24. Krason, W., Wysocki, J.: Investigation of friction in dual leaf spring. J. Frict. Wear 38(3), 214–220 (2017). https://doi.org/10.3103/S1068366617030096

25. Konopka, S., Łopatka, M.J., Spadło, K.: Analiza mobilności kołowych platform przegubowych. Logistyka-nauka, Nr 6, 5633–5641 (2014)

26. STANAG 2021: Wojskowe obliczenia klasyfikacji mostów, promów, tratw i pojazdów, Wydanie 6

Investigation of Helicopter Impact on Structural Vibrations of Elevated Helipads

Wiesław Krzymień[(✉)] [ID]

ŁRN - Institute of Aviation, Warsaw, Poland
`wieslaw.krzymien@ilot.edu.pl`

Abstract. Elevated helipads at hospitals may have a diverse structure and location- depending on the possibilities of the hospital and its environment. The Vibroacoustics Laboratory of the Institute of Aviation performed the measurement of the vibration properties of several helipads of varying degrees of construction. The purpose of the research was to determine the vibration properties of the elevated helipads and to try to estimate the impact of a helicopter's landing and taking-off on the building on which it was placed. This paper presents some results of measurements including comparing measurements made during landing and take-off of the helicopter and those excited with an impulse, carried out for two elevated helipads. The goal was to estimate (model) the impact of the helicopter on a helipad construction at the stage of its design and construction in the range of low frequency (up to 100 Hz).

Keywords: Elevated helipads · Ground vibration tests · Measurement of vibrations

1 Introduction

The air transport system for victims of accidents or seriously ill patients facilitates the provision of prompt and specialized medical assistance. The two main solutions of the hospital helipads are:

- ground helipad requiring large space,
- elevated helipads most often used in built-up areas (e.g. near the city center).

In Poland there are currently registered more than 250 hospitals helipads, and among them there are 36 elevated helipads (Fig. 1) which operate 24–h a day [1].

Due to the requirements and area of operation of the Polish Medical Air Rescue, (LPR) the number of hospitals helipads is expected to rise. Most of them are built above hospital buildings.

An elevated helipad occupies a small area and also allows to shorten the route between the helicopter and a Hospital Emergency Department (SOR). The disadvantages are associated with e.g. high construction costs, safe operation requirements, high winter

A. Mężyk et al. (Eds.): SMWM 2020, AISC 1336, pp. 181–190, 2021.
https://doi.org/10.1007/978-3-030-68455-6_16

maintenance costs and the possibility of dangerous consequences of a helicopter accident during take-off or landing.

A separate problem is the impact of a helicopter's landing and taking-off on the hospital, its patients, staff and equipment. The Regulation of the Minister of Health in force since 1 July 2019 [2] requires that an elevated helipads should be designed so that it does not affect the functioning of the hospital building and facilities through the impact of vibrations and noise.

Fig. 1. Helicopter EC135 during the first landing on the elevated helipad on hospital in the city center of Warsaw.

Placing the elevated helipads on the building causes a direct impact of a landing and taking-off helicopter, which is an important source of vibration and noise [3]. The impact on the immediate surroundings during landing and shutting down the propulsion and the subsequent start-up and take-off is caused by:

– engine noise, noise of the rotor blades (tips) and tail propeller noise,
– pulsation of the blow from the rotor blades,
– load from a "hard landing",
– vibration of the main rotor transmitted to the landing gear.

The main source of excitation of the helicopter's vibrations is the main rotor. The pulsation of the airflow from the blades under the rotor also occurs when the helicopter does not have direct contact with the elevated helipads (shortly before landing or immediately after detachment) while the range of frequencies generated is associated with the rotation of the carrier rotor and the number of blades. Air ambulance helicopters EC-135 may be a source of vibration with a nominal frequency of rotor operation (approximately 6.5–6.8 Hz depending on version) and multiples thereof: number of blades (4) and number of blades ± 1 (i.e. 3 and 5). During the start-up or braking of the helicopter's main rotor due to the changing rotation frequencies, there may be short-term vibration excitations may occur resulting from the transition through different resonance frequencies of the structure (i.e. 0 to 34 Hz). Starting the engine takes approximately 1 min, stopping approx. 1.5 min, so the passages through the resonances of the helicopter-helipad system are fast.

Elevated helipads can be either of concrete or steel construction (Fig. 2) depending on a building's structural strength and possibility of performance.

An important feature of elevated helipads currently required by Regulation of the Min. of Health is the so-called "airgap" - the space between the roof of the building and the board of the helipad, which in case of stronger wind stabilizes the flow of the building and the helipad increasing the safety of landing and take-off of the helicopter [4].

Recently introduced standards PN-B-02170_2016 [5] and PN-B-02171_2017 [6] determine the measurement method and recommended vibration levels on the floors of different rooms in a building. The required range of measured vibrations includes frequencies between 1 and 100 Hz.

Fig. 2. Part of a helipad construction of a hospital in Olsztyn.

Designing an elevated helipad on a hospital building requires calculations and vibration analysis of the structure in terms of both the structure and the helicopter's impact on the building [7]. Another important element is also knowledge of the vibration properties of the building itself.

The actual impact of the helicopter and measurements of the vibration level in the building can only be carried out during its take-off and landing operations, however their practical implementation involves a temporary disruption of the hospital's activities as sensors need to be located in places which may not be accessible to the unauthorized persons (e.g. operating rooms). A separate threat to the building, patients and hospital staff are emergency conditions, such as helicopter's hitting on the landing board or turbulence caused by strong winds.

Measuring the effect of the helicopter's impact forces on the elevated helipads during take-off or landing is technically difficult to perform: the impact is not local, and any special construction (e.g. sensor platform) will affect the results through its dynamic properties. In addition, full measurement should cover all 6 degrees of freedom (component forces and moments). In this situation, it was decided to estimate the frequencies and magnitudes of the vibrations caused by the helicopter by comparing the object's response to another local impact, which is impulse inducing. This approach allows estimating the forces while carrying out simple and quick measurements of the structure at any stage of the helipad construction. The results allows to verify the accepted dynamic properties of the helipad and building as well as to introduce possible structural changes.

Preliminary measurements of the acoustic vibrations impact on the EC-135 environment are shown in [8]. This article presents and compares sample results of measurements of the vibrations of the elevated helipads and the building raised by the helicopter and vibration raised by the impulse of force.

2   Measurement Method

Studies of the vibration properties of the helipads and upper floors of buildings enabled to estimate the shape and frequencies of vibrations that can occur. Measurements were made in three stages:

– assigning the shape of free vibration of helipad,
– assessing the transferability of vibrations from the helipad board to the floor of the lower floors of the building [9],
– registering vibrations at selected construction points during the landing and take-off of the helicopter.

The tests were conducted by analysing the signal from the sensors after the object was excited with an impulse generated by a modal hammer of 5.6 kg with a soft tip and, for comparison, a 30 kg sandbag (Fig. 3). Figure 4a. shows a graph of the force flow for four different strokes (force impulses), from which it appears that the impulse time does not depend much on the impact force and excited vibrations were up to approx. 150 Hz. Similar measurements were made for the sandbag: several discharges were made to assess the force impulse at a special scales: the plate under which the force sensors were mounted. Oscillations with a frequency of approx. 200 Hz (Fig. 4b) were associated with free vibrations of this plate. The use of a sandbag to excite vibrations of the helipad board was dictated by:

- a sufficiently large mass for a person to perform a repeatable excitation - discharge from a height of up to about 1.5 m,
- adequate (assumed) excitation energy,
- avoiding damage of the helipad board,
- simplicity of conducting the measurements.

Fig. 3. Modal hammer and sandbag on the helipad.

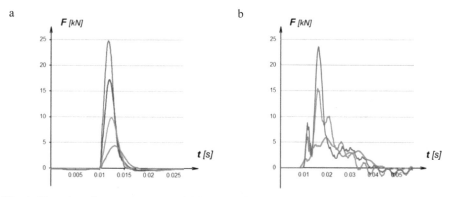

Fig. 4. Force impulse courses: a - modal hammer (for 4 different stroke), b - sandbag (dropped from a high of 0.5, 1 and 1.5 m).

Table 1 juxtaposes basic key parameters of impulse excitation with the hammer and the sandbag.

It is assumed that the typical volume of energy transferred during the landing corresponds to a free fall of the helicopter from a height of around 2 cm. In the case of the EC135 helicopter (weight of approx. 3000 kg), the energy transferred would correspond

Table 1. Key parameters.

Quantity	Hammer	Sandbag
Mass [kg]	5.6	30
Time of impulse [ms]	~4.5	~25
Max force [kN]	~40	~25
Impulse of force [Ns]	<90	~160
Energy [J]	<700	~440

to approximately 580 J. The limit value defined by the regulations for the construction of helicopters (e.g. CS 27.725) is the so-called "discharge test" which involves discharging the helicopter (free drop) from a height of 13" (33 cm) to estimate the strength of the chassis. For the EC135 helicopter, the transferred energy is approx. 9700 J.

Measurements of vertical vibrations were carried out using sensors glued in the middle of the elevated helipads structure - as shown in Fig. 5. Sensors No. 1 and 2 were glued symmetrically from the underside of the plate near the center of the elevated helipads, while sensors No. 4, 5, 10 and 11 at the base of the central pillars on both sides of the insulation.

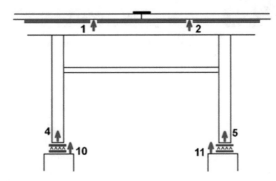

Fig. 5. Distribution of sensors near the center of the helipad.

LMS measurement equipment (SCADAS) and dedicated software was used for research.

3 Results

The real loads of the helipad due to a 'hard' landing are not known and difficult to register: the pilots perfectly know how to land gently even in difficult conditions, especially with patients on board. The vibration levels recorded during the two consecutive landings in good weather turned out to be very similar. Figure 6 shows an example time recording from sensors No. 1 and 2 during the landing and take-off of the helicopter (it was placed not in the center of the helipad but near sensor No. 2.

a

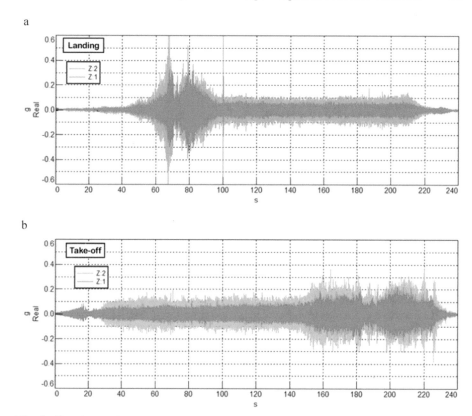

b

Fig. 6. Signals recorded by sensors No. 1 and 2 registered during landing (a) and take-off (b) of the helicopter.

Figure 7 shows signals recorded by sensors No. 1 and 2 after impulse excitation by the modal hammer and the sandbag. The levels of excited vibration are comparable.

a   b

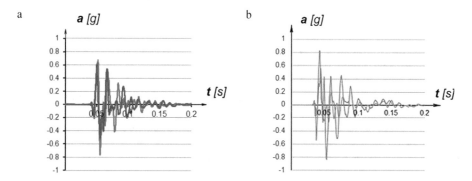

Fig. 7. Signals recorded by sensors No. 1 and 2 registered during impulse excitation made with: a - modal hammer (force approx. 4 kN), b - sandbag (dropped from a height of about 0.5 m).

Figure 8 shows the amplitude-frequency characteristics obtained from the signal recorded by the sensors after impulse excitation in the center of the helipad board.

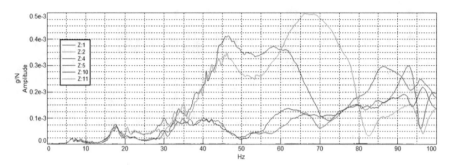

Fig. 8. Amplitude-frequency characteristics obtained from the recording of the signal from the sensors after impulse excitation in the center of the helipad board.

Signals recorded on concrete pillar bases (10 and 11) in Fig. 8 are barely visible, which means that vibrations transferred to the lower floors are negligible due to effective vibration insulation.

A simplified wavelet analysis can be used to compare the effectiveness of impulse excitation and vibrations caused by landing and take-off of the helicopter. The analysis (based on the weave of functions) consisted in multiplying the fragments of signals with a large amplitude recorded during the landing and take-off of the helicopter by a shortened response received after hitting the modal hammer as well as the sandbag (Fig. 7). The obtained values were normalised, i.e. divided by the value obtained from multiplication of the wave signal by itself. It was assumed that for the frequency range covered by PN-B-02171_2017 standard [6] (i.e. 1–100 Hz) it was sufficient to spread the signal of 1600 samples/s. Part of the signal from Fig. 6 with the highest amplitude and including the touchdown of the helicopter (Fig. 9a) was filtered and reduced by the number of samples. The graph of this signal is shown in Fig. 9b. Reducing the amplitude boundary of registered accelerant sensor signals indicates that a large part of the vibrations had frequencies above 200 Hz.

Figures 9c and 9d represent the results of the wavelet signal analysis (e.g. [10]) from sensors No. 1 and 2 with a signal after impulse excitation with a hammer and a sandbag: the increase in the vibration amplitude with frequencies generated by impulse with flight descent and rotor operation is visible. In second 78 a response is visible after the helicopter's skid hit the helipad (according to [8] and video recording: the left skid hit the helipad near sensor No. 2). Based on the received values, it was estimated that the impact was very small and corresponded to an impulse excited by a hammer with a force of ca. 2 kN or discharging of a sandbag from a height of ca. 20 cm, which corresponds to the energy of only about 50 J. These results are similar to the results of helicopter force landing calculations shown in [11].

The analysis carried out for the helicopter landing and take-off from another helipad gave similar results.

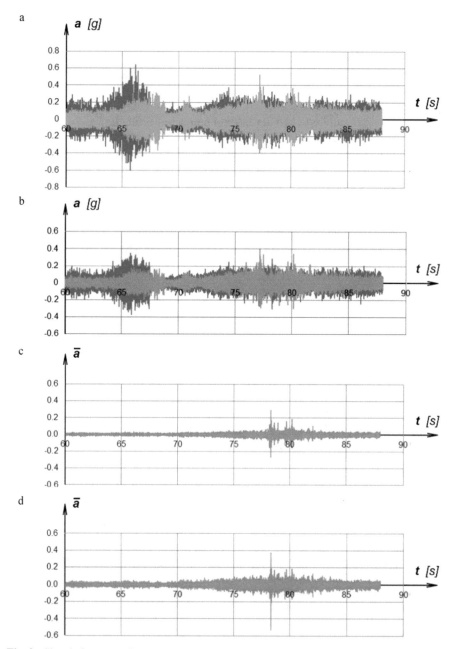

Fig. 9. Signals fragments from sensors 1 and 2 recorded during helicopter landing: a - a fragment of the signal records from Fig. 6, b - the same fragment after reducing the number of samples, c - results of the wavelet analysis of the signal excited with a modal hammer, d – results of the wavelet analysis of the signal excited with a sandbag.

4   Conclusions

1. The conducted measurements allowed to obtain information about the vibration properties of elevated helipads. The impulse excitation shown in the article is a simple and sufficiently effective way to assess the vibration properties of the helipad at any stage of its construction.
2. The helicopter's full impact on the elevated helipads is difficult to measure. A direct comparison of the magnitude of vibrations generated during the landing and take-off of the helicopter and by excitation of impulse with comparable energy allows to assume that this impact is sufficiently similar in terms of frequency and magnitude of excited vibrations.
3. The final, although difficult to perform, verification of the helicopter's impact on the construction of the elevated helipad and the hospital building is the analysis of the signal from multiple sensors during the helicopter landing and take-off.
4. An effective but costly method of assessing the impact of a helicopter on the helipad and the building is constant monitoring of vibrations at selected points of the helipad and the building.

References

1. LPR Homepage: https://www.lpr.com.pl/en/
2. Regulation of the Min. of Health of 27.06.2019 on the hospital emergency department (J. of Law. 2019 pos. 213)
3. EASA Certification Specification CS-27 Small Rotorcraft
4. Federal Aviation Administration, US Department of Transportation, 2012, Heliport Design -AC 150/5390–2c, Chapter 4-Hospital Heliports
5. Polish Standard PN-B-02170_2016 "Ocena szkodliwości drgań przekazywanych przez podłoże na budynki" (Assessment of the harmfulness of vibrations transmitted by the earthen foundation to buildings)
6. Polish Standard PN-B-02171_2017 "Ocena wpływu drgań na ludzi w budynkach" (Assessment of the impact of vibrations on people in buildings)
7. Wąchalski, K.: Assessment of the current construction conditions for elevated helipad on hospital buildings in Poland. Trans. Inst. Aviat. No. 3(244), 189–201 (2016)
8. Cieślak, S., Krzymień, W.: Initial analysis of helicopter impact on hospital helipads. Trans. Aerospace Res. 256, 14–23 (2019)
9. Krzymień, W., Cieślak, S.: Investigation of the Vibration Properties of Concrete Elevated Hospital Helipads. Vibrations in Physical Systems No 31 (2020)
10. Edwards T.: Discrete Wavelets Transform: Theory and Implementation. Stanford University (1991)
11. Stanisławski, J.: A simulation model for computing the loads generated at landing site during helicopter take-off or landing operation

The Use of Alginate Mass to Obtain Limb Geometry for the Sake of Human and Animal Limb Prosthesis

Tomasz Machoczek![ORCID] and Agnieszka Konopelska[(✉)]![ORCID]

Department of Theoretical and Applied Mechanics, Silesian University of Technology, Gliwice, Poland
agnieszka.konopelska@polsl.pl

Abstract. The aim of this study is a practical analysis of the innovative yet non-invasive use of a method obtaining complex human and animal limb geometry on the basis of human and animal limb with its main goal to develop a personalized prosthesis. The hereby article presents an overview of widely known and used external evaluation methods as well as inner limb stump geometry for the sake of fabrication of upper and lower limb prosthesis. A different approach has been presented to limb geometry modeling which is featured by a great shortening of time with a patient due to the use of alginate mass. Because of a lack of possibilities to get limb stump geometry, in the experiment a healthy human and animal limbs were used. The methodology to get a complex human and animal limb shape, presented in this article, involves in the first place the realization of negative cast with the use of alginate mass and secondly positive cast with the use of plaster. The following stages of work include healthy limb 3D scanning and 3D printing in order to make a real-life model.

Keywords: Limb prosthesis · Residual limb · Limb stump · Negative/PositiveCast · 3D scanning · 3D printing

1 Introduction

When designing a limb prosthesis it is essential to create an adequate shape of a limb stump. In order to carry out a precise measurement, different methods are applied to get the limb stump geometry. Before getting down to design the limb it is important to assess both inner and outer limb geometry. Moreover, an essential element to design a limb prosthesis is an assessment of biomechanical properties of residual limb soft tissue. To assess the outer geometry of a limb stump the following methods can be applied–water immersion and circumferential measurements. In both cases the measurement of residual limb volume is carried out. In various types of contact methods a measurement of the surface contour of a residual limb by scanning of negative/positive cast or directly the residual limb. Among other methods used to assess limb stump geometry we can enumerate Moirecontourography, laser video scanning. Silhouetting method as well as hand-held scanner. On the other hand, internal geometric assessment is done with the use

A. Mężyk et al. (Eds.): SMWM 2020, AISC 1336, pp. 191–200, 2021.
https://doi.org/10.1007/978-3-030-68455-6_17

of medical imaging: X-ray imaging, computed tomography (CT), spiral X-ray computed tomography (SXCT), magnetic resonance imaging (MRI) and ultrasound imaging. The assessment of biomechanical properties of limb soft tissue concerns mainly indentation measurement, vibration measurement and measurements of material properties [2, 4, 15–17].

Presently applied contact and contactless methods of obtaining contours of damaged limbs are realized by means of traditional techniques with the use of measuring rules and through modern 3D scanning techniques. Unfortunately, each of them has got many drawbacks. In case of measurement by means of different measuring elements, in view of geometrical complexity shown by limb stumps (as well as complete limbs) operation of engineering hypersurface imaging which they represent is constricted and time-consuming. Unfortunately it can be stressful for the patient who is required to keep his limb still and motionless, quite often for a longer period of time.

Methods which allow indirect shaping of limb crater cause great hardship with implementation of its remaining components. Limb creation based on that technique is carried out on the patient almost within the full range of time essential to develop the construction.

The use of scanning for geometry imaging requires stabilizing the limb which is a great challenge for a man whereas it can turn out insufficient and difficult to do with an animal without narcotizing or shaving it. Furthermore, scanning almost always requires taking sequential photos which make up for the unreadable contours which happened as a result of limb motion.

On the other hand, joining the above mentioned techniques and a trial to create so called hybrid method transfers to a total time of prosthesis development despite a shortened to minimum contact with a patient. That's why nowadays newer and newer materials are used to build a form on the basis of which a model of a limb stump can be made. Consequently on its basis through a visual technique (optical methods) we can obtain a surface virtual model for the design of geometrically optimal prosthesis with a possibility to conduct simple and quick corrections and improvements without the presence of a patient.

2   Methods of Obtaining Geometry of a Limb Stump

Browsing literature one can find different types of techniques to obtain the shape of a limb stump belonging to a person after amputation for the sake of stump cast. The authors of the thesis present the suspension casting technique developed at Northwestern University. Such a technique allows to obtain a stump cast in conditions of controlled suspension. The authors of thesis [3] developed the limb stump casting method to which they used a special casting stand.

Another method of limb creation is 3D laser scanning system [7, 11]. In this method contactless 3D scanning enables monitoring of changes of lowerlimb'smorphology. Another method developed by the authors of the thesis [1] is based on prosthesiscrater forming with the use of a pressurized casting sack. The use of plaster of Paris (POP) bandages and plaster has been presented in [6]. The method is based on wrapping a limb stump with anorthopedic bandage like plaster of Paris and marking and modifying the pressure relief are aatpatellartendon in negative cast. The method of using alginate mass has been presented in this article by the authors of thesis [10]. The Drape Forming method is based on the use of flat, heated thermoplasticsheet to form prosthesis socket [8]. The Art of Moulding a prosthetic socket is based on forming prosthetic socket by means of bubble method presented broadly in thesis [9]. The authors of thesis [12] suggested using computed tomography (CT) to image a limb stump with the use of mirrored version of a healthy limb. In Fig. 1 there is a set of methods presented in this article about how to obtain a limb shape or limb stump. Table 1 presents pros and cons of various methods of limb or stump limb geometry sourcing.

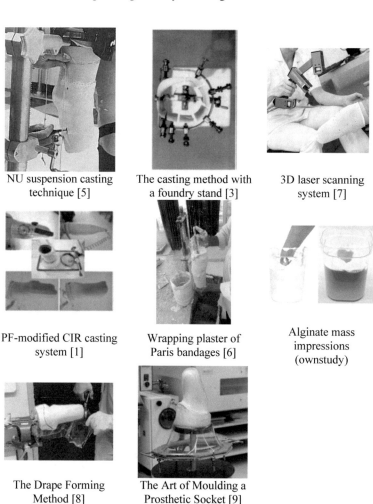

NU suspension casting technique [5]

The casting method with a foundry stand [3]

3D laser scanning system [7]

PF-modified CIR casting system [1]

Wrapping plaster of Paris bandages [6]

Alginate mass impressions (ownstudy)

The Drape Forming Method [8]

The Art of Moulding a Prosthetic Socket [9]

Fig. 1. Methods of limb orstump limb geometry sourcing.

Table 1. Pros and cons of various methods of limb orstump limb geometry sourcing.

Method of limb orstump limb geometry sourcing	Mapping quality	Requirements related to patient sleep	Geometric compliance	Process complexity	Restrictions (application)	Lead time	Stressogenicity
NU suspension casting technique	proper	may require (animal)	proper	complex	mainly used in humans	2 h<	high
The casting method with a foundry stand	adequate	may require (animal)	adequate	complex	mainly the lower limb	2 h+	high
3D laser scanning system	proper	may re-quire (ani-mal)	proper	medium complex	without limits		high
PF-modified CIR casting system	proper	may re-quire (ani-mal)	very good	complex	used for conical-shaped limb stumps		high
Wrapping plaster of Paris bandages	proper	may require (animal)	proper	complex	mainly used in humans	2 h<	high
Alginate mass impressions	excellent	not required	excellent	not complex	anywhere on the body	2 h<	low
The drape forming method	proper	may require (animal)	proper	complex	mainly used in humans	2 h<	high
The art of moulding a prosthetic socket	proper	may require (animal)	proper	complex	mainly used in humans	2 h<	high

3 Material and Method of Casting the Human and Dog Limb

Focusing on a model being a replica of a human hand we can choose from a range of standard products out of which one can differ in price, complexity and speed of mixing, time of bonding, various hardness after bonding, measurement stability as well as its colour and change while bonding including smell. Making a choice between available mass and operational time and costs one can acknowledge as a front-runner of the task a product called Kromopan 100 by Lascod company. The material has the longest bonding time, the lowest price and a favourable time to make a mould. In this case bonding time of 3 min is not a key parameter, and taking into account that the parameter is given for fixed working conditions which do not include such factors like hand temperature causing the acceleration of bonding process.

After the material has been chosen for the form it is essential to think about the purpose of it. In view of an assumed high quality of model imaging it has been beneficial to reach for dental plaster which shows good fraction abilities and efficient pouring with interesting colours and waterproofing features as well as available hardness rate. In the project a high quality plaster of 4 rate of hardness has been used called Model typo Pink by Lascod which in the moment of bonding transfers from pink to delicate brown (natural skin colour), proving the completion of total bonding.

Stages of forming negative cast have been presented on an example of a hand. In order to obtain a negative cast of a hand defined actions have been carried out in the following order: a timer was switched on which controlled the time of mass mixing for the form. After that a big vessel was filled with cold water (temperature definitely below 23* C in order to slow down the reaction process) of volume increased in 5% to the value appointed above for Kromopan 100 material. It was supposed to increase the pouring of material. The following stage was to put into a vessel with water a desired mass destined for a mould and then a dynamic mixing was started by means of a mixer. After about 45 s mixed mass was poured into a vessel in which a mould was to be made and afterwards for 5 s one hit the vessel onto a worktop with a rolled towel placed on it. Such action aimed at smoothing alginate mass with total lack of air bubbles. Into such prepared mass a cool hand was submerged which was previously moistened with water. When the colour of the mass turned from pink to white, about 1 more minute was given to bond the total volume and as soon as the mass congealed, the hand was carefully taken out by moving it in various directions which allowed to unstuck it from the surface of the form. The time from the moment of putting the hand to alginate mass till it was taken out was about 105 s. Due to a short time of mass action it can also be used with animals. To compare, authors of thesis [10], in which alginate mass was also used to have a negative cast of lower limb stump, state the time as much longer because up to 5 min which in case of animals would require narcotizing or premedication. Figure 2 presents the Kromopan 100 material used in the experiment plus stages of negative cast.

Fig. 2. Stages of forming a negative cast on the example of the hand.

For the earlier appointed volume of plaster one prepared delicately diluted mass of the material which was immersed in the mould remaining after hand removal. At this point it is advisable to have a bigger amount of plaster because it is possible to achieve a hand base. Such a step requires the repetition to shake the inside of the vessel in order to remove the air bubbles and the same time filling the form completely. Action which brings a beneficial effect in that issue is filling the form with pancake batter mass directly down the wall. After 3 h a form with plaster mould was removed from the vessel. Next, by pushing fingers onto a form one outlined soft areas void of plaster mould. From that point taking off the material was launched. Such action can be realized manually or with a knife and the spots full of details, due to the plaster fragility, it is more efficient to clean by means of a wooden toothpick. It is worth considering that as long as the mould is wet in full volume is easy to separate from the model but when it dries it gets hard and difficult to remove. If, for any reason, we want to postpone in time the process of mould isolation then it is good to shut tightly the vessel with both mould and cast and put it into the fridge. The final action (however not obligatory to introduce) concerns the removal of a base, namely so called intentional or unwanted rising. In such a case it is necessary to break off manually or cut off the excessive material using a small kitchen knife. Figure 3 presents the stages of moulding a plaster cast. The sight of cast straight after rising removal is presented in Fig. 4 (hand) and Fig. 5 (front dog limb).

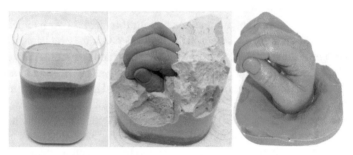

Fig. 3. Pouring the mold with gypsum mass and stages of removing the mold from the plaster model on the example of a hand.

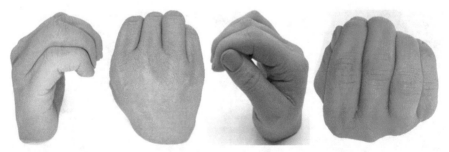

Fig. 4. View of the hand plaster cast just after removing the riser.

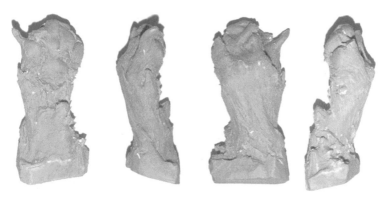

Fig. 5. View of the plaster cast of the dog's paw just after removing the riser.

4   3D Scanning and Printing of Limb

After having a positive cast of human hand 3D scanning and printing was applied in order to create a real-life model. In hereby article it was presented how efficient the conducted trials were for sourcing the complex geometry for chosen layout of healthy limbs in the shape of: partly clenched woman's hand, man's stretched hand and a hairy dog's paw. A great attention was put, first of all, to the hand with fingers stretched, for which a shell model was developed in terms of additive technology production with the use of modern production methods. All the models have a common feature which is the time to obtain a cast. However they differ in time needed to develop shell models which gets longer depending on what irregularities in shape they represent. It is directly connected with the necessity to develop more scanning pictures essential to build a complete cloud of points describing the analyzed geometry. Figure 6 presents a cast of a hand with upright fingers.

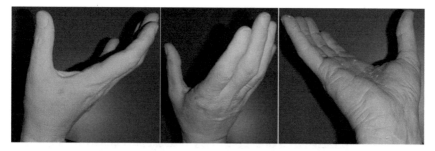

Fig. 6. A plaster cast of a hand with straight fingers.

Scanning of hand plaster cast involved the three following stages. At first a shell model was created which was obtained as a result of scanning with the use of structural light laser (settled in the work area of a tool to generate G-code). The second stage was oriented to obtain shell model with contour lining on the base of a determined height and the level of filling. The last stage of 3D scanning is to obtain a model which has a standoff structure for the sake of prototyping. Scanned hand plaster cast enabled the creation of a real model with the use of additive techniques based on a method of fused deposition modeling [FDM]. Figure 7 presents the stages of shell model creation whereas Fig. 8 presents a 3D printed human hand.

Fig. 7.   Stages of creating a shell model.

Fig. 8.   3D printed human hand.

5   Summary

Nowadays computer supported designing and fabrication of human and animal limb stumps, to a great extent, can make work easier for the orthotists. In the thesis [4, 13, 14, 18] the application of Computer- Aided Design/Computer Aided Manufacture (CAD/CAM) and Finite Element Anaysis (FEA) for the sake of limb prosthesis fabrication has been presented. In this article presented methodology of obtaining complex limb shape or human and animal lib stump can contribute to reducing the time of prosthesis fabrication as well as to minimizing stress of an animal during the negative cast.

The application of low-budget alginate mass allocated, to a greater extent, to man ufacture small prosthetic cast can be an efficient tool to obtain geometry of human and animal limbs of a bigger volume. An asset which undoubtedly is a quick time of mass crystallization but at the same time long enough to get a high quality cast which in turn reduces the stress of a patient and eliminates the need to narcotize the patient. The possibility to manipulate the bonding time which is obtained by local lowering of the limb temperature, the application of water below 20 *C as a bonding substance as well as the use of various types of chemical reaction retardents like Time outs can allow to adjust freely to unusual placement and geometry of an analysed limb. The lack of chemical interaction between petroleum jelly and chromatic alginate allows to use this method safely even with hairy patients without any need of additional depilation which contributes negatively to stress and the overall operation time. Furthermore, the visually signaled state of alginate mass bonding time, thanks to colour changes, allows to estimate correctly the model what, in addition, makes it easier to develop a model for a target prosthesis.

References

1. Jivacate, T., Devakula, M.L.B., Tipaya, B., Yesuwarn, T.: Twenty-one months' experience with the PF-modified CIR casting system for trans-tibial prostheses. Prosthet. Orthot. Int. **35**(1), 70–75 (2011)
2. Zheng, Y., Mak, A., Leung, A.: State-of-the-art methods for geometric and biomechanical assessments of residual limbs: a review. J. Rehab. Res. Develop. **38**(5), 487–504 (2001)
3. Gleave, J.A.E.: A plastic socket and stump casting technique for above-knee prostheses. J. Bone Joint Surgery British 47(1), 100–103 (1965)
4. Colombo, G., Filippi, S., Rizzi, C., Rotini, F.: A new design paradigm for the development of custom-fit soft sockets for lower limb prostheses. Comput. Ind. **61**(6), 513–523 (2010)
5. Hampton, F.: NU suspension casting technique. Bulletin of Prosthetics Research. Fall. pp. 52–65 (1966)
6. Nayak, C., Singh, A., Chaudhary, H.: Customised prosthetic socket fabrication using 3D scanning and printing. Conference paper: Additive Manufacturing Society of India (2014)
7. Science Photo Library Homepage. https://www.sciencephoto.com/media/464417/view/orthopaedic-and-prosthetic-3d-scanning Accessed 5 Oct. 2020
8. Orfit Homepage. https://www.orfit.com/blog/drape-forming-method/ Accessed 5 Oct. 2020
9. Orfit Homepage. https://www.orfit.com/blog/moulding-a-prosthetic-socket/ Accessed 5 Oct. 2020
10. Engsberg, J.R., Sprouse, S.W., Uhrich,, M.L., Ziegler B.R., Luitjohan, F.D.: Comparison of Rectified and Unrectified Sockets for Transtibial Amputees. J. Prosthet. Orthot.. **18**(1), 1–7 (2008)
11. Colombo, G., Bertetti, M., Bonacini, D., Magrassi, G.: Reverse engineering and rapidprototyping techniques to innovate prosthesis socket design. Three-Dimensional Image Capture and Applications VII. Proceedings Volume 6056 (2006)
12. Cabibihan, J.J., Abubasha, M.K., Thakor, N.: A method for 3D printing patient-specific prosthetic arms with high accuracy shape and size. IEEE Access. **6** (2018)
13. Sewell, P., Noroozi, S., Vinney, J., Andrews, S.: Developments in the trans-tibial prosthetic socket fitting process: a review of past and present research. Prosthet. Orthot. Int. **24**, 97–107 (2000)
14. Colombo, G., Facoetti, G., Morotti, R., Rizzi, C.: Physically based modelling and simulation to innovate socket design. Comput.-Aided Des. Appl. **8**(4), 617–631 (2011)
15. Seminati, E., Talamas, D.C., Young, M., Martin Twiste, M., VimalDhokia, V., Bilzon, J.L.J.: Validity and reliability of a novel 3D scanner for assessment of the shape and volume of amputees' residual limb models. Public Library of Science. September 8(2017)

16. Hajiaghaei, B., Ebrahimi, I., Kamyab, M., Saeedi, H., Jalali, M.: A comparison between the dimensions of positive transtibial residual limb molds prepared by air pressure casting and weight-bearing casting methods. Med. J. Islamic Rep. Iran **30**, 341 (2016)
17. Krouskop, T.A., Dougherty, D., Yalcinkaya, M.I., Muilenberg, A.: Measuring the shape and volume of anabove-kneestump. Prosthet. Ortho. Int. **12**, 136–142 (1988)
18. Silver-Thorn, M.B., Childress D.S.: Generic, geometric finite element analysis of the transtibial residual limb and prosthetic socket. J. Rehab. Res Develop. **34**(2) (1997)

Intelligent Drive in Industry 4.0 – Protection of Toothed Belt Transmission on the Basis of Its Digital Twin

Julian Malaka[✉] [iD] and Mariusz Hetmańczyk[iD]

Silesian University of Technology, Konarskiego 18A, 44-100 Gliwice, Poland
`julian.malaka@polsl.pl`

Abstract. In modern industry, the autonomy in responding to threats is a highly valued feature of machines. The ability to quickly collect and process data on the condition of equipment and events in its environment is the basis for many security or alarm solutions. However, some components are excluded from ongoing monitoring due to difficulties in placing and operating sensors in their area. This state of affairs causes diagnostic methods to be developed in the direction of inferring about given damage with limited access to signals clearly related to specific symptoms. One of the elements of Industry 4.0 – a digital twin – may be helpful in this area. The advantages of such a solution can be observed on the example of virtual representation of the toothed belt drive. The publication describes a proposal to use a dynamic model of a device for the induction of information about the behaviour of a flexible transmitting element only on the basis of the recording of the input and output motions in the mechanism under consideration. The transmission is treated as a "black box", which can be accessed through a computer simulation. This allows one to determine the condition of the toothed belt, which strength translates into the reliability of the machine. The knowledge gained in this way plays a key role in the system of automatic protection against the breaking of the belt. The principles of the intelligent drive and the result of its experimental verification are presented in the publication.

Keywords: Industry 4.0 · Digital twin · Expert system · Belt transmission · Drive technology

1 Introduction

Machines – both technological and transport ones – as well as a huge part of mechanisms used in everyday life are often equipped with toothed belt transmissions. They constitute an ideal solution when there is a need to connect in a specific way a generator and a receiver, being distant from each other, or to adjust the output characteristic of motion. The devices under consideration, being in general use, have been structurally refined over many years. There are many geometric and dimensional variations available in standardised series of types. Materials and the machining the parts of the systems of this kind are relatively inexpensive, and the technical parameters meet most of the drive needs. Moreover, the rotational motion can be easily converted in to the translational motion. Toothed belt transmissions feature the synchronisation of passive and active pulley rotations, which is essential in many cases. It is used wherever precise position control is required [1–4]. Nowadays, the objective is to reduce the expenditure of time and money spent on routine – sometimes unnecessary – maintenance procedures. Their number can be minimised with appropriate protective measures. Market trends suggest the development of this area towards automation, computer techniques and innovative diagnostic methods. They are of particular importance in the era of lean manufacturing and the fourth industrial revolution.

The publication presents an idea of the automatic diagnostics of a drive equipped with an elastic toothed belt, using a solution based on a numerical simulation with regard to mechanics and data flow and processing within the framework of the Industrial Internet of Things and machine learning. As a result of the considerations, there was proposed a system in which a digital twin of the transmission provides on an ongoing basis detailed information pertaining to the behaviour of the belt when there occur given conditions determined by several easily measurable parameters of an actual object. Thanks to that, there occurs the reasoning about the operating state of a device on the basis of the premises in the nature of recorded and modelled signals. It leads to the formulation of decisions being essential in the appropriate adjustment of a technological process. The availability of digital data processing and the possibilities it offers make the information solutions be increasingly desired in engineering and are the subject of numerous research in this field (Fig. 1).

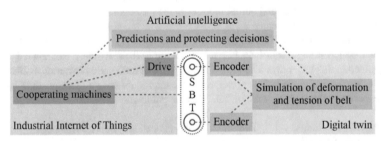

Fig. 1. Diagram of proposed protection system, where SBT – synchronous belt transmission.

The main objective of the activities undertaken is the automation of the operations under consideration, so that the system equipped with the tool being developed could operate in an "intelligent" way. The notion is understood as the creation of knowledge by gathering and analysing experiences, prediction of the consequences of identified events, and the resulting adaptive control. The researches included an attempt to integrate the environments in which it is possible to perform individual tasks essential for the comprehensive implementation of the authors' original concepts. It was proposed to apply popular standard measures, developed and used widely in manufacturing engineering for many years. The CAE (Computer Aided Engineering) software has become the basic tool for mechanics, designers, integrators and installers. It allows them to create various forms of the visualisation of the design and operation of machines on the screen or in the virtual or augmented reality environment. It was preliminarily checked whether the selected means enable the information processing and flow in a way that would result in the efficient autonomous identification of risks and taking appropriate measures in line with the adopted assumptions. This kind of knowledge determines the implementation potential of the solution being designed in terms of the protection against damage to drive components. The considerations gave rise to the formulation of further research perspectives, and provided the basis for the indication of the novel application of techniques which are the pillars of Industry 4.0.

2   Methods

Within the research, an experimental verification of the validity of the proposed concepts was conducted. What is nowadays most commonly used to model mechanisms in engineering is the software based on the multi-body system (MBS). Its purpose is to solve differential equations of motion – dynamic ones, in the case under consideration – enabling the simulation of the reactions of individual component bodies of the analysed device to given excitations and constraints. This allows one to determine the forces acting on a specific element in given circumstances, which can constitute the basis for conclusions about its overload and risk of damage. For the visco-elastic component under consideration, the Kelvin-Voight model was used, with a division of the body into many identical rigid bodies connected with dampers and springs. The effectiveness of this technique was investigated and presented in many scientific publications, including [5–8] (Fig. 2).

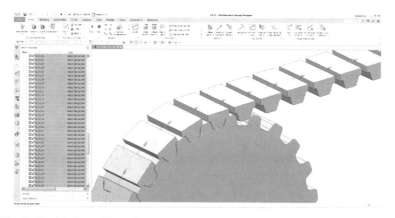

Fig. 2. Partial view of investigated model in environment for described experiments.

As part of the conducted work, an attempt was made to implement the resulting digital replica of the device in the adopted diagnostic system. It was assumed that demonstrating the possibility of it interacting in a certain way with appropriate IT tools and thus generating the desired signals is the confirmation of the authors' vision. The obtained results constitute the basis for the evaluation of the potential and the indication of the directions for the development of the tools which were the subject matter of the study described in the publication. The tasks were performed in an integrated computer aided engineering environment Siemens NX, using the Mechatronics Concept Designer module, the functionality of which meets the recognised research needs.

2.1   Concept of Digital Twin Auto Validation

The exact course of the simulated quantities depends not only on the quality (the correctness of the implementation) of the model, but also on the degree of the overlapping of its parameters with the characteristic of the original. The selection thereof is a subject matter of the validation, i.e. the determination whether a given digital twin match the specific case being analysed within the desired range [9, 10]. The authors propose to automate this task with the application of the additional equipping of the tested transmission with sensors for partial registration of the belt behaviour. What is investigated is the measurements of the displacement of its certain parts or the form of vibrations, which can be identified by non-contact methods. This way, the features of the observed object would not be interfered with, it would not be necessary to install any metrological subassemblies on the surface of the belt and to ensure communication with them, which, in many cases, is virtually impossible (Fig. 3).

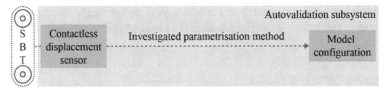

Fig. 3.   Signal flow in concept of autovalidation process.

The acquired information would be a signal to the learning system, indicating a possible discrepancy between the simulation and real events, and thus a need to make corrections to the calculations. The research conducted so far has allowed the authors' to estimate the chances of implementing the described solution in terms of its autonomous validation. The latter is conditioned by the result of the analysis of the possibilities of integrating particular elements within the framework of the tool being designed.

2.2   Concept of Integrating Real Object with Its Numerical Replica - Simulation Control

The coupling elements for the real and virtual object are to be the angular positions of pulleys. These parameters are the input information in the simulation – the excitations defining the operating circumstances of the analysed critical part of the transmission. The elastic belt reacts to the motion of rigid bodies with which it is engaged through the contact conditions specified in the model. Registering the required quantities in the physical system is not problematic (as opposed to obtaining accurate data on the deformation of the transmission component). Rotation monitoring is commonly applied, with popular sensors such as encoders providing very high accuracy when compared to other solutions. In many cases, they are pre-installed on drives to, among others, implement various motion control techniques. The signals which are indispensable in the proposed system can therefore be received from the controller of a machine and, if necessary, a device can be additionally equipped with a small and easy to install measuring set (Fig. 4).

Fig. 4. Signal flow in concept of subsystems integration.

The practical value of the proposed concept of integrating the model with the real object was evaluated through an attempt to establish connections between various production and simulation data processing environments.

2.3   Concepts for Automatic Reasoning on Operational Status, Using Digital Twin

The accurate mapping of the belt dynamics allows one to omit any means of recording its behaviour and base the analysis solely on a theoretical scenario. It is created as a result of inferring from the premises in the form of a fragmented description of the situation in the transmission (provided in the process of the simulation control) and the quantities modelled. Thanks to the suggested approach, information about the condition of an elastic component in given moments and circumstances is to be artificially generated; it is to be used to formulate predictions or to trigger safeguards when necessary. It has a significant impact on the strength of the observed subassembly and the maintenance of the entire production line.

Apart from the cases resulting from the ongoing drive monitoring, the learning system could automatically perform additional trial calculations with different sets of configuration parameters and excitations. The experiments would be aimed at creating a base of knowledge on the causes, course and consequences of specific events. Gathering the resulting information would lead to an increase in the speed and accuracy of the response of the digital twin to signals from the physical object. Matching the existing situation to the pattern which was discovered earlier in the described manner would mean an immediate diagnosis and undertaking an appropriate reaction without time-consuming mathematical operations. Access to such data would allow the resulting scenarios to also take into account the probability of the occurrence of phenomena not implemented in the baseline model, such as the absence of the belt (caused e.g. by its break) or change of its features (Fig. 5).

The proposed concepts are feasible if the simulated values of the interactions in the transmission can be observed on an ongoing basis and sent to the appropriate tools for processing the information thus obtained. The assumed functionality was evaluated.

Fig. 5. Diagram of proposed intelligent system for data generation within digital twin.

3   Results

The undertaken considerations and attempts led to the creation of a prototype data exchange system, which generates and signals conclusions about the condition of the tested elastic element. All necessary modelled parameters are updated in keeping with the course of the simulation events. The access to them is the basis for the formulation of a procedure for identifying risks and taking preventive actions. As early as at this level it is possible to indicate relationships and the correctness in the course of signals, helpful in the implementation of control and partial identification of specific states (Fig. 6).

Fig. 6. Definition of rules for automatic belt state identification.

This ensures that alarming and necessary safeguards are activated. In the investigated system, the response to the set conditions (relations between the rotation of the pulleys) was generated immediately (Fig. 7).

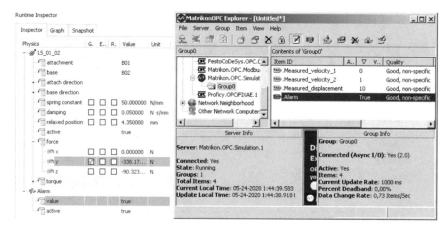

Fig. 7. Alarm indication after occurrence of dangerous state of belt.

The Mechatronics Concept Designer module enables the input and output of data, using, among others, the OPC (Object linking and embedding for Process Control) standard. In the solution being designed there was established the communication between the modelling environment and the industrial server (Fig. 8).

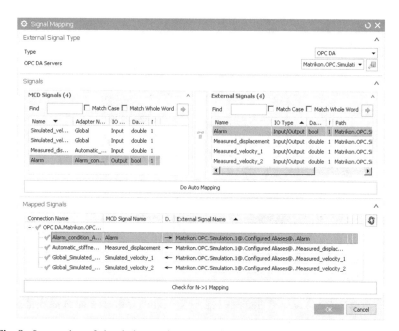

Fig. 8. Integration of simulation environment with universal industrial collector of data.

The characteristic of virtual objects is adjusted on an ongoing basis as a result of applying the parametric configuration thereof (Fig. 9).

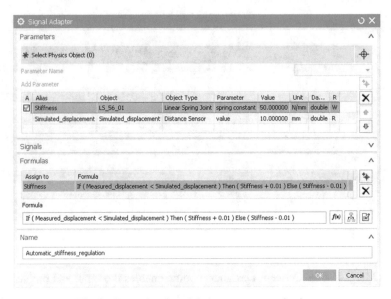

Fig. 9. Example of model element parametrisation.

Both production field devices and other software can be the source and receiver nodes in the developed information flow network (Fig. 10).

Fig. 10. Preparation of created digital twin for cooperation with external numerical system.

In the Siemens NX there was also implemented special support for the interaction of the simulation with the MATLAB application [11]. It was used to prepare the channel of data exchange with this computational tool.

4 Discussion and Conclusions

Industrial controllers and other field devices are designed for the OPC communication. This makes the used server a central node, where all numerical values relevant to the system being developed can be saved and updated on an ongoing basis. This way, the process data is made available to the physical equipment and computer applications connected in the proposed way. The aim is for them to exchange information and use it autonomously according to their needs, in line with established rules. Thus, the entire concept constitutes a solution under the Internet of Things standard. The value of the parameter sought in the created network may come from many sources, the redundancy of which is the basis for the execution of the auto-validation task, according to the described concept. The integration of various environments for generating and processing signals and the assumed flow thereof – essential for the target functionality of the tool being designed – are therefore feasible. This is the foundation of the desired automatic configuration or control operations. The verified interaction of the Siemens NX software and MATLAB constitutes the confirmation of the capabilities and the preparation of the system for the development towards advanced inference and prediction using artificial intelligence methods.

The premises, which consist of external signals (ultimately coming from industrial equipment) and modelled quantities, are the basis of the inferring about what the cause of the observed phenomena is – i.e. how the toothed belt behaves – and what it leads to. The calculations therefore produce the information necessary to create a set of examples on the basis of which the learning system could create patterns for recognising a risk.

4.1 Further Research Perspectives

The research proved that the application of the digital twin can constitute a solution to the indicated diagnostic problems related to the toothed belt transmissions. The available measures of process automation make the behaviour of physical objects be hypothetically determined with computer models. The presented vision is in line with Industry 4.0 standards and has great potential to become a vital element of predictive maintenance. Nevertheless, the described technique requires further development towards model-based inference and the autonomy of machines when comes to taking appropriate control or alarm actions. The development of the system which learns, gathers and processes symptoms, and formulates appropriate indications will be the next milestone in the work being conducted. It is believed that an appropriate tool may be created using the MATLAB computing environment. It is used for many similar tasks and enables the integration with the Siemens NX software, by means of which the discoveries described herein were made. The target solution would make it possible to activate security measures not only within a single device, but also to harmonize all phases of the production cycle in terms of safety and minimisation of wear and tear of the analysed component.

References

1. Srivastava, H., Haque, I.: A review on belt and chain continuously variable transmissions (CVT): Dynamics and control. Mech. Mach. Theory **44**, 19–41 (2009)

2. Almeida, A., Greenberg, S.: Technology assessment: energy-efficient belt transmissions. Energ. Build. **22**, 245–253 (1995)
3. Pakuła, A.: Wpływ wybranych czynników na trwałość pasów zębatych napędu rozrządu silnika spalinowego. Archiwum Motoryzacji 4 (2005)
4. Malaka, J., Wróbel, A., Płaczek, M.: The network control system of high-bay warehouse. MATEC Web of Conferences **112**(05004), 1–6 (2017)
5. Callegari, M., Cannella, F.: Lumped parameter model of timing belt transmissions. In: 5th AIMETA Congress of Theoretical and Applied Mechanics, Taormina, Italy, Sept. (2001)
6. Čepon, G., Manin, L., Boltežar, M.: Validation of a flexible multibody belt-drive model. J. Mech. Eng. **57**, 7–8 (2011)
7. Čepon, G., Manin, L., Boltežar, M.: Introduction of damping into the flexible multibody belt-drive model: A numerical and experimental investigation. J. Sound Vibr. **324**(1–2), 283–296 (2009)
8. Kubas, K.: A model for analysing the dynamics of belt transmissions with a 5pk belt. The Archives of Automotive Engineering **67**(1) (2015)
9. Balci, O.: Verification, validation, and certification of modeling and simulation applications. In: Proceedings of the 2003 Winter Simulation Conference, pp. 150–158 (2003)
10. Karkula, M.: Weryfikacja i walidacja dynamicznych modeli symulacyjnych procesów logistycznych. Logistyka **2**, 717–726 (2012)
11. Siemens NX software documentation https://docs.plm.automation.siemens.com/tdoc/nx/12.0.2/nx_help Accessed 24 May 2020

Assessment Method of Machinery Technical Safety in the Aspect of Virtual Commissioning of Machines

Andrzej Melecki◉ and Piotr Michalski(✉) ◉

Faculty of Mechanical Engineering, Institute of Engineering Processes Automation and Integrated Manufacturing Systems, Silesian University of Technology, ul. Konarskiego 18 A, 44-100 Gliwice, Poland
piotr.michalski@polsl.pl

Abstract. New technologies used for construction and upgrading formulate new research tasks. That is why the consideration of new solutions, methods and acquisition of new information in this respect is an important aspect which contributes to the progress of knowledge in this area. This paper applies to one of the issues, namely the assessment method of machinery technical safety. It presents the essential aspects related to legislation, risk assessment and analysis and selection of the risk reduction model. Solving of the problem is important because the difficulty occurring in the process results in ambiguous assessments and may generate additional hazards in the system, causing a reduction in the safety level. On the stage of assessment and reduction of the risks which occur in the mechatronic system, one can refer to the formal and legal area, which describes the harmonisation mechanism, but the relevant types and methods used for conducting a hazard analysis, as well as risk assessment tools, can be selected in any way, depending on the system safety engineer's expertise. The authors recommend that a uniform assessment method of the machinery technical safety should be adopted, which will help to model cohesive virtual models developed for the needs of virtual technical commissioning.

Keywords: Virtual commissioning of machines · Technical safety · Risk assessment · Risk reduction

1 Introduction

The issues related to an attempt to describe an assessment method of machinery technical safety have been presented in the Polish and global scientific literature for over twenty years [1–4]. On the one hand new technologies have provided gradually more effective and universal machines but on the other hand new, previously unknown problems have emerged in the area of a systemic and effective assessment procedure of this issue. The problems concerning the issue can be considered in the following areas: a formal and legal analysis of the relevant requirements, based on the existing information resources; selection of the risk assessment model; selection of the hazard analysis type, depending

© The Author(s), under exclusive license to Springer Nature Switzerland AG 2021
A. Mężyk et al. (Eds.): SMWM 2020, AISC 1336, pp. 211–222, 2021.
https://doi.org/10.1007/978-3-030-68455-6_19

on the system life cycle phase; correlation of the adequate hazard analysis methods with the selected type; determination of the risk assessment tools; adopting a model determining the acceptable risk value and selection of the risk reduction model. Other important aspects include: over dimensioning of the control system architecture in the part of functional safety measures implementation; neural networks; Internet, including but not limited to the IoT (Internet of Things) and IIoT (Industrial Internet of Things); fuzzy logic methods; self-programming and self-configuring systems; voice control of machines and cybersecurity of technological production systems. These technologies will formulate new research tasks, and the aforementioned challenges, in the light of a lack of non-ambiguous methods, determine the directions of looking for solutions, both in the areas of assessment methods formulation and completion as well as safety systems designing, programming and validation. That is why the consideration of new solutions, methods and acquisition of new information in this respect is an important element of these activities and contributes to general progress in the knowledge of this area. This paper concerns one of these issues, which is the assessment method of machinery technical safety. It includes the aspects related to the description of the technical safety assessment method in relation to legislation, risk assessment and analysis and selection of the risk reduction model. Although on the stage of analysis, risk assessment and reduction of the risk occurring in the mechatronic system one can refer to the formal and legal area describing the mechanism of harmonisation, the selection of the right types and methods used for hazard analysis as well as risk estimation tools can be carried out in any way, depending on the system safety engineer's expertise.

2   Technical Safety Assessment Method

As a result of the research, a technical safety assessment method was developed, which contributes to an extension of the description and systematics, and general progress in the method development.

A general formalism of the approach to the assessment of the machinery technical safety was adopted, based on a structure in which its adequate levels can be differentiated. The first stage includes a formal and legal analysis of the adequate requirements concerning machinery technical safety, based on the existing information resources. The next stage is characterized by the selection of the risk assessment method. Then, depending on the system development cycle phase, the right type of hazard identification is applied, and then the correlation is carried out between the adequate analysis method and the previously applied type. Finally, the procedure leads to the evaluation of the risks existing in the system and to the determination of the right method of risk reduction. The following stages of the machinery technical safety assessment procedure have been identified: formal and legal analysis of the adequate requirements based on the existing information resources, selection of the risk assessment model, selection of the hazard analysis type depending on the system life cycle phase, correlation between the adequate hazard analysis method and the selected type, determination of the risk assessment tool, adopting a model determining the acceptable risk value, selection of the risk reduction model.

The following subsections describe the substantial content of each step of the procedure.

2.1 Formal and Legal Analysis of the Respective Requirements, Based on the Existing Information Resources

The first stage of the adopted procedure for developing a technical safety assessment method covers an analysis of the adequate requirements based on the applicable regulations. The information was classified in three basic areas: social directives, economic directives (technical harmonisation) and other requirements. The execution of the relevant provisions causes an assumption that the requirements are met. The scope of the subjects of analysis is shown in Table 1.

Table 1. Valid regulations.

Social directives	Economic directives (technical harmonisation)	Other requirements
89/391/EEC, 89/654/EEC, 2009/104/EEC, 89/656/EEC, 90/269/EEC, 90/270/EEC, 2004/37/EC, 2000/54/EC, 92/57/EEC, 92/58/EEC, 92/85/EEC, 92/91/EEC, 92/104/EEC, 93/103/EEC, 98/24/EC, 1999/92/EC, 2002/44/EC, 2003/10/EC, 2013/35/UE, 2006/25/EC	89/392/EEC, 91/368/EEC, 93/44/EEG, 93/68/EEC, 95/16/EC, 98/37/EC, 98/79/EC, 2006/42/EC	Acts of law which transpose the harmonising EU legislation into domestic legislation: acts, regulations and declarations. Executive acts: regulations, declarations, orders, resolutions, decisions, reports, judgements of the Court of Justice, announcements. Corporate requirements. Harmonised standards

The following existing information resources were identified as part of this stage: thematic materials available on the websites devoted to legislative regulations, directives in the technical harmonisation area, social directives, and regulations [5].

2.2 Selection of the Risk Assessment Model

Another stage of the adopted procedure for the development of the technical safety assessment method is related to the selection of the risk assessment model. Based on the analysis of the formal and legal documents, a model was selected fulfilling the provisions of the harmonised EN ISO 12100 [6]. The risk assessment in this algorithm includes risk analysis and evaluation. Risk analysis involves collection, recognition and processing of the available information, necessary to identify hazards and estimate the risk.

Risk evaluation completes the process of assessment. The risk analysis provides the information necessary for its evaluation, which in turn enables making decisions about the need or no need to reduce the risk. An assessment should include studies referring to all stages of the machine life, such as: construction, transport, and commissioning, i.e. assembly, installation and adjustments that affect the safety; operation - setting, learning, normal work, cleaning, troubleshooting, maintenance and repairs; withdrawal from use, disassembly and disposal as scrap. The flow of the risk assessment process, with reference to detailed provisions of the harmonised EN ISO 12100, is shown in Fig. 1.

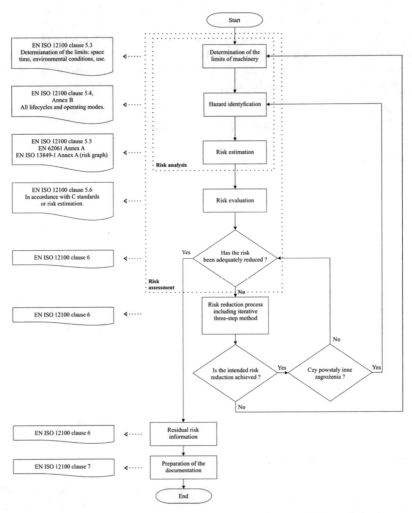

Fig. 1. Risk assessment flow chart with reference to detailed provisions of the harmonised standard [6].

The following step shall include the selection of the hazard analysis type, depending on the system life cycle stage.

2.3 Selection of the Hazard Analysis Type Depending on the System Life Cycle Stage

On the stage of risk assessment model selection, the harmonisation mechanism refers to EN ISO 12100, indicating a number of logical steps, but the adequate types and methods used for conducting a hazard analysis can be selected freely, depending on the systems safety engineer's expertise. The analyses are intended for a systematic examination of the system, subsystem, object, components, software and staff and their mutual links. There are two main categories in the system, referring to the analysis type and the method of its performance. Each type of analysis is supposed to provide adequate tools, depending on the system life cycle phase, which will easily identify the hazards occurring on this stage of the system development. The methods shall be carefully selected in order to reach measurable objectives of each analysis type [7]. In the system safety area there are the following seven basic types of hazard analysis:

- CD-HAT - Conceptual Design Hazard Analysis Type
- PD-HAT - Preliminary Design Hazard Analysis Type
- DD-HAT - Detailed Design Hazard Analysis Type
- SD-HAT - System Design Hazard Analysis Type
- OD-HAT - Operations Design Hazard Analysis Type
- HD-HAT - Health Design Hazard Analysis Type
- RD-HAT - Requirements Design Hazard Analysis Type

Figure 2 shows the concept of a hazards filter in the system development cycle on the design stage, using the basic types of hazard analyses. In this concept, each model of hazard analysis acts like a screen identifying certain kinds of hazards which occur on the particular stage of the product life cycle. Every subsequent filter is used for the identification of new hazards that were not identified on the previous level of the audit. Bold, dark arrows on the top of the filter stand for hazards occurring at the beginning of the system development. After all types of hazard analysis have been applied, only the known hazards are left, reduced to the permissible risk level. They are marked with smaller and thinner arrows, representing the residual risk. The use of all seven types of hazard analysis is of key importance for identification and mitigation of all hazards and reduction of the system residual risk. Figure 3 shows hazard analysis types in the whole life cycle of the system. Each type of analysis is supposed to provide adequate tools, depending on the system life cycle stage, which easily identify hazards on this stage of the project development. This in turn opens up an opportunity to acquire more detailed information concerning a specific type of hazard analysis, because more detailed information about the project and operation becomes available as the system develops.

The detail level of the hazard analysis increases as more project details become available. Each type of analysis defines the moment when an assessment should start, the level of the analysis detail, type of information available and the outcome of the analysis.

The important rule which applies to the hazard analysis is that one specific type of analysis does not necessarily identify all hazards and cause-and-effect factors in the system, and so seven different models were assumed. The type of analysis determines where, when and what shall be analysed, establishes specific analytical tasks of the particular system life cycle stage, describes what is required in the analysis, and puts emphasis on the design stage.

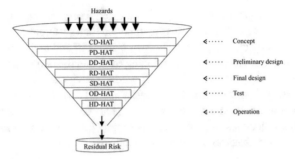

Fig. 2. Hazard filer in the system development cycle on the design stage [7].

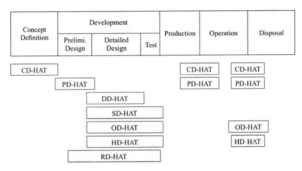

Fig. 3. Types of hazard analysis in the whole life cycle of the system [7].

2.4 Correlation of the Adequate Hazard Analysis Methods with a Selected Type of Analysis

The most common analysis methods which are used by experts working in the area of systems safety include: PHL (Preliminary Hazard List Analysis), PHA (Preliminary Hazard Analysis), SSHA (Subsystem Hazard Analysis), SHA (System Hazard Analysis), O&SHA (Operating and Support Hazard Analysis), HHA (Health Hazard Assessment), SRCA (Safety Requirements/Criteria Analysis), FTA (Fault Tree Analysis), ETA (Event Tree Analysis), FMEA (Failure Mode and Effects Analysis), Fault Hazard Analysis, Functional Hazard Analysis, SCA (Sneak Circuit Analysis), PNA (Petri Net Analysis), MA (Markov Analysis), Barrier analysis, BPA (Ben Pin Analysis), HAZOP analysis, CCA (Cause Consequence Analysis), CCFA (Common Cause Failure Analysis), MORT analysis, SWSA (Software Safety Assessment). The correlation between the adequate hazard analysis method and the selected type is included in Table 2.

Seven out of over 22 most common methods are regarded as the primary methods used by system safety practitioners. The methods include: PHL, PHA, SSHA, SHA, O&SHA, HHA, and SRCA. The Preliminary Hazard List (PHL) is developed on the stage of conceptual design, initial design or on the system commissioning stage, and it is a starting point for future hazard analyses, which may turn out to be indispensable for the identification of all hazards in the system. The Preliminary Hazard Analysis is an

Table 2. Methods used in each type of hazard analysis [7].

Type of analysis	Method of analysis
CD-HAT	PHL (Preliminary Hazard List Analysis)
PD-HAT	PHA (Preliminary Hazard Analysis)
DD-HAT	SSHA (Subsystem Hazard Analysis), FTA (Fault Tree Analysis), FMEA (Failure Mode and Effects Analysis), FaHA (Fault Hazard Analysis), FuHA (Functional Hazard Analysis), SCA (Sneak Circuit Analysis), PNA (Petri Net Analysis), MA (Markov Analysis), BPA (Ben pin analysis), HAZOP (Hazard and Operability Studies), CCA (Cause Consequence Analysis), CCFA (Common Cause Failure Analysis), MORT (Management Oversight Risk Tree Analysis), SWSA (Software Safety Assessment)
SD-HAT	SHA (System Hazard Analysis), FTA, ETA (Event Tree Analysis), FuHA, SCA, PNA, MA, BA (Barrier Analysis), HAZOP, CCA, CCFA, MORT, SWSA
OD-HAT	O&SHA (Operating and Support Hazard Analysis)
HD-HAT	HHA (Health Hazard Assessment)
RD-HAT	SRCA (Safety Requirements/Criteria Analysis)

inductive, non-standardised method used for identification of hazards, hazardous situations and near misses in all stages of the system, subsystem or element life cycle. The purpose of the PHA analysis is to carry out a risk assessment, taking into consideration the severity of the potential effects, which consequently translates onto planning of preventive and risk reduction measures. This means proposing adequate safety measures and the result of their application. The main stages of the PHA include: initial findings (determination of the purpose and scope, selecting the team, collecting the information etc.), hazard identification, estimation of the loss occurrence probability and the severity of the effects according to the assumed scale, risk ranking and the subsequent preventive measures. The process of analysis can be carried out as a "brainstorm", involving a systematic review of the available design documentation. A table is developed based on the collected information, indicating the estimated probability of the effects occurrence, their severity and risk. The risk ranking is meant to determine whether the risk is acceptable or non-acceptable. The hazard assessment carried out on the stage of the initial design or commissioning of the machine creates an opportunity to identify the hazard areas, which helps to predict specific preventive measures already on the early stage of the design or commissioning. The results of the PHA can be used for comparing different design concepts or as a contribution to more detailed risk analysis. The description of the results can be presented in different ways (e.g. as a chart, logic tree or in a numerical form).

2.5 Determination of the Risk Estimation Tool

After hazard identification, the possibility of the potential user's contact with the hazard shall be determined alongside with the risk it could cause (body injuries, material and financial losses, other). The accident risk is a stochastic notion. Hence the risk estimation involves its calculation with a certain degree of probability on the required level of confidence. A risk is a combination of the loss occurrence probability and its severity [6].

A loss is a body injury or deterioration of health. The qualitative character of risk estimation determines the individualised approach to each analysis, where expertise, history of accidents with the same or similar machines, statistical studies etc. shall be applied. There are no standard risk estimation methods. From the point of view of a research method, there are two basic types of risk analysis: deductive and inductive, while with regard to the character of the result, the risk estimation methods can be divided into qualitative and quantitative. There are many numerical assessment tools used for risk estimation with the numerical score method. The possibility of selecting one integer, within specific classes, can provide a wider choice than in qualitative categories, but it may also give a false impression of numerical accuracy. The numerical assessment tools have two or more parameters, divided into several classes, similarly to risk matrixes and risk graphs. The difference is that relevant numerical values rather than quality notions are assigned to risk classes. In order to obtain a numerical score of the estimated risk, a class is selected for each parameter, and the related values or results are then combined, e.g. by adding or multiplication.

$$PHR = f(PO, FE, PA, DPH) \tag{1}$$

$$PHR = PO \cdot FE \cdot PA \cdot DPH \tag{2}$$

$$0,0046875 \leq PHR \leq 9750 \tag{3}$$

The numerical scoring systems enable easy and non-ambiguous weighing of the parameters.

The developed model of technical safety assessment has adopted the PHR number as the risk estimation tool.

PHR (Pilz Hazard Rating) is the function of a hazardous incident PO (Possibility of Occurrence), FE (Frequency and Duration of Exposure), PA (Possibility of Avoidance) and DPH (Degree of Possible Harm). This relationship is described by function (1). PHR is the product of relevant numerical values of the risk elements, and it is calculated based on relationship (2). Table 3 summarizes the values that the risk elements may take in relation to the PHR number. The calculated coefficient value can range from 0.0046875 to 9750, where 0.0046875 means a lack of risk, while 9750 stands for the highest risk possible. This correlation is described by relationship (3).

2.6 Adopting a Model Determining the Acceptable Risk Value

The proposed approach specifies six levels of risk: negligible risk, very low risk, low risk, significant risk, high risk and very high risk. The achievable risk levels were also determined. They were correlated with the PHR number values and the proposed risk reduction measures.

Table 4 summarises the achievable risk levels. Following the risk assessment, the risk should be evaluated in order to determine whether its reduction is necessary. The

Table 3. Risk elements of the PHR number [2].

Risk element	Value	Description
PO	0.05	Almost unlikely
PO	1.25	Highly unlikely
PO	2.5	Possible
PO	4	Probable
PO	6	Certain
FE	0.5	Once a year
FE	1	Every month
FE	2	Every week
FE	3	Every day
FE	4	Every hour
FE	5	Continuous
PA	0.75	Possible
PA	2.5	Possible under certain conditions
PA	5	Impossible
DPH	0.25	Scratching or bruising
DPH	0.5	Cut or wound or a mild effect
DPH	3	Minor fracture (finger or toe)
DPH	5	Major fracture (hand, shoulder or leg)
DPH	8	Loss of one or more fingers
DPH	11	Arm or leg amputation, partial loss of vision or hearing loss
DPH	15	Amputation of both arms, complete loss of vision or hearing loss
DPH	25	Hazard to life or permanent loss of health
DPH	40	Fatal (one casualty)
DPH	65	Disaster (several casualties)

evaluation means a decision to accept or reject the assessments concerning the estimated risk levels which will be acquired in the risk analysis. Should it be necessary to reduce the risk, adequate protective measures shall be selected and used. The iterative process also involves checking if the implementation of the protective measures did not contribute to the occurrence of additional hazards or an increase in another risk. If new hazards have occurred, they shall be included in the list of identified hazards, and adequate protective measures shall be recognised for them.

Table 4. Risk levels correlated with the PHR number [2].

PHR	Risk level	Description
0.0046875–10	Negligible risk	Almost no risk to health and safety, further risk reduction measures are not required
11–20	Very low risk	Very low risk to health and safety, significant risk reduction measures are not required. The use of PPE or staff training is required
21–45	Low risk	Risk to health and safety is present but low. The implementation of risk reduction measures shall be taken into consideration
46–160	Significant risk	Substantial risk which requires the implementation of risk reduction measures at the next suitable opportunity
161–500	High risk	Potentially dangerous hazard, which requires immediate implementation of risk reduction measures
501–9750	Very high risk	Risk reduction measures shall be implemented immediately. The information shall be communicated to the corporate management

2.7 Selecting the Risk Reduction Model

Risk assessment is followed by risk reduction. The process iteration may be necessary on this stage to eliminate the hazards to a practicable degree. The risk is reduced by implementing protective measures which are a combination of measures used by the designer and the user. The measures that can be applied on the design stage are preferred over those applied by the user, and they tend to be more effective. The maximum practicable risk reduction is the objective. The objective can be fulfilled by eliminating the hazards or by eliminating - separately or simultaneously - each of the two risk elements: severity of the loss caused by the referenced hazard and the probability of the loss occurrence.

Once the risk reduction objectives have been achieved, one can be convinced that the risk has been reduced sufficiently. In the adopted model, it is achieved with a three-step method. The method is shown schematically in Fig. 4, including references to the detailed provisions of PN-EN ISO 12100.

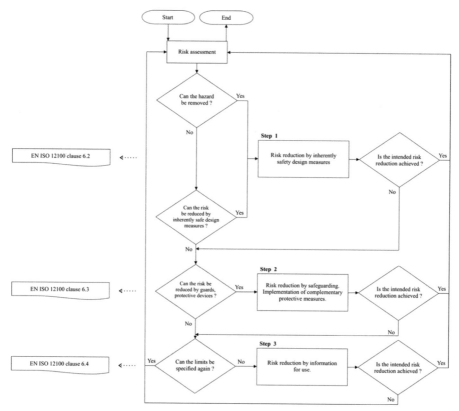

Fig. 4. Three-step method with reference to the provisions of PN EN 12100 [6].

3 Summary

There are several basic ways to ensure the safety of the machine use. First and foremost, the number of occurring hazards and near misses shall be reduced by implementing intrinsically safe construction solutions. Then the stay of humans or access shall be reduced as much as reasonably practicable. After all these possibilities have been exhausted, other concepts, including functional safety measures shall be implemented.

A technical safety assessment method was developed as a result of the research carried out, which helped to extend the description and systematics, and contributed to a general progress in the method development. The final result - diagram of the detailed safety assessment procedure - is shown in Fig. 5.

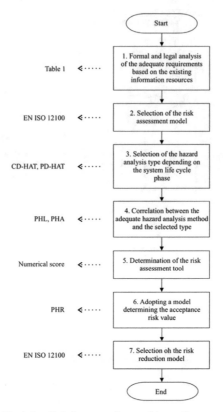

Fig. 5. Final detailed diagram of a machine safety assessment.

References

1. Macdonald, D.: Practical Machinery Safety. Elsevier, Amsterdam (2004)
2. Pilz Gmbh&Co.: Pilz: Guide to Machinery Safety (2004)
3. Stephans, R.: System safety for the 21st century. Wiley, Hoboken (2000)
4. Ridley, J., Pearce, D.: Safety with Machinery. BH-Elsevier, Amsterdam (2006)
5. EUR-lex. https://www.eur-lex.europa.eu
6. EN ISO 12100 Safety of machinery – General principles for design – Risk assessment and risk reduction (2012)
7. Cliford, A.: Hazard Analysis Techniques for System Safety, 2nd edn. Wiley, Hoboken (2015)

Testing the Motion Resistance of Angular Contact Spindle Bearings for Different Ways of Preload Implementation

Mateusz Muszyński$^{(\boxtimes)}$ ⓘ and Andrzej Sokołowski ⓘ

Department of Machine Technology, Silesian University of Technology, Konarskiego 18A St., 44-100 Gliwice, Poland
mateusz.muszynski@polsl.pl

Abstract. The paper presents the results of experimental studies of motion resistance in angular contact ball bearings used in electrospindles of machine tools. The motion resistance directly impacts the amount of heat generated during operation, which is related to thermal deformations and machining accuracy. Development of a theoretical model of motion resistance would enable machine tool designers to optimize the design of electrospindles. In order to develop such a model, it is necessary, among others, to adjust the coefficients of friction of the theoretical model to the results of experimental research. The paper focuses on assessment of influence of preload and rotational speed on motion resistance. In practice, the angular bearings are generally preloaded in two ways, namely using elastic elements, e.g. disc springs, or without elastic elements. The elastic preload refers to spindles that work at particularly high rotational speed, while the rigid preload is a conventional solution when it comes to spindles of machine tools. Tests of motion resistance were performed for both types of preload by introducing slight structural modifications at a test rig.

Keywords: Angular contact ball bearing · Motion resistance · High speed cutting

1 Introduction

Contemporary spindles of machine tools are faced with very high kinematic requirements (permissible rotational speeds of several tens of thousands rpm), with increasing machining accuracy. There is a need to use adequate bearing in order to meet such high requirements. Today, high-speed spindles are commonly used with angular contact ball bearings [1]. They are applied because they can achieve very high speed with a relatively low motion resistance. Also, they offer possibility to transfer load axially and radially at the same time, easier assembly and operation. Furthermore, the angular bearings make the construction of bearing nodes significantly simpler.

The adequate operation of angular bearings requires the so-called preload, namely the initial axial force that loads a pair of bearings. The way of applying this load is important from the perspective of bearing stiffness or the motion resistance torque. Spindle angular

A. Mężyk et al. (Eds.): SMWM 2020, AISC 1336, pp. 223–233, 2021.
https://doi.org/10.1007/978-3-030-68455-6_20

contact bearings are commonly loaded in one of two ways. Most often there are two sleeves grinded to reach an appropriate dimension and placed between a pair of bearings. In this way, after the nut is tightened there is no clearance between the bearing ring and the sleeve (Fig. 1a) [3]. In the literature, this type of preload is called rigid preload or fixed position preload. The value of the load applied in this manner is determined by the displacement of the internal ring in relation to the external one, expressed in μm. The stiffness of such nodes is high, but as the rotational speed increases, particularly at the maximum speeds, there is an increase in motion resistance arising from high centrifugal forces and spin torque [2, 6].

In the case of spindles operating at very high speed (milling centers, grinders), elastic elements such as disc springs are used to preload the bearings (Fig. 1b) [3]. The application of an elastic element causes that the bearing is loaded with a constant force of a specific value. As the rotational speed and centrifugal forces increase in the bearing, there are some slight axial displacements between the rings, which slightly change the disc spring tension force. It is assumed that for this type of preload the tension force is constant and independent from the rotational speed.

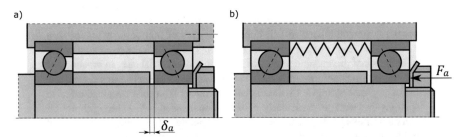

Fig. 1. Ways of applying preload of angular spindle bearings: a) rigid preload, b) elastic preload.

The resultant force of preload is transferred between the inner ring, balls and the outer ring. The reaction force between the ball and the raceway is called contact force. One of the analytical models that allows estimating the resistance torque resulting from rolling friction is the Musiał model [8] (dependence 1). This model requires prior determination of contact forces. Analytical determination of the contact forces is complex and time-consuming, but the model is still under development in order to consider the elastic radial deformations of a rotating ring [9] or thermal deformations.

$$M_{1_{(T)}} = \left(\frac{d_m}{D} + 0,5\right) \sum_{j=0}^{j=Z-1} Q_j f_{kj} \qquad (1)$$

where:

d_m - bearing pitch diameter,
D - ball diameter,
Z - number of balls,
Q_j - jth substitute contact force,
f_{kj} - jth rolling friction coefficient.

Apart from the rolling friction, the total torque of motion resistance also includes viscous resistance resulting from the presence of grease and the resistance torque which is the effect of the spinning phenomenon [2]. To this day, the moment of viscous resistance is easily determined using Palmgren's dependence [11]. In turn, the spin phenomenon consists in the fact that each ball makes a resultant rotational motion around its own axis, perpendicular to the contact zone between a ball and a raceway. This motion results from non-parallelism of the ball's rotation axis during its rolling and the bearing axis [2]. Such a motion of the balls generates additional friction, which is of great significance in the case of high speeds.

Analytical modelling of motion resistances requires numerous calculations which are time-consuming [2, 6]. An easier way to estimate the motion resistance is experimental testing. The following part of this paper presents the concept of a test rig allowing for the implementation of preload for both methods mentioned above. Section 2 of the paper is related to all construction solutions applied in the rig. Section 3 presents measuring setup, methods for recording and processing of data together with some exemplary results. In turn, Sect. 4 shows the results of research on the influence of rotational speed and preload on motion resistance as well as the conclusions from the conducted tests.

2   Test Rig

The first concept of a test rig assumed that the bearing will be preloaded elastically with disc springs. Figure 2a presents the design of such a rig. The B-7007-E-T-P4S bearing pair in the "O" arrangement is loaded with a nut 12. Between the nut and the inner ring clamping element 13, there is a piezoelectric force sensor 6 (Kistler type 9102A) for measurement of the preload force. There are four disc springs located between the bearings, to increase the spring package susceptibility and facilitate the loading. Due to the quite unusual external dimension of the bearing ($\phi = 62$ mm), it was necessary to use an additional thin-walled sleeve to centre the spring package in the shaft 3 rotation axis. The length of the above-mentioned sleeve was chosen so that the springs could properly deflect. The shaft and its bearings are mounted to an aluminum housing 2 bolted to the base 4. The body has holes to measure the temperature of the bearing outer rings during long-term operation. The whole structure is screwed to piezoelectric torque sensor 5 (Kistler type 9272). The bearings are mounted on both the shaft and in the housing with sliding fit, so that both can be preloaded and relative axial displacement of the rings can take place during operation. The shaft is connected with a flexible jaw clutch to the Tecnomotor electrospindle, which is controlled by the Sinamics V20 module allowing stepless speed control.

All measurements that were carried out concern the operation of two bearings lubricated with NBU 15 plastic grease.

In view of unreliable preliminary results obtained, it was decided to modify the rig. The modification consisted in the replacement of disc springs, center sleeve and spacer sleeves with a single sleeve, which was based on both sides of the bearing outer rings. The elastic element was a rubber washer placed between the piezoelectric sensor 6 and the pressure element 13. The high compressibility of the rubber made it easier to determine the exact amount of preload - many turns of nut 12 are needed to obtain a given preload value. Due to the very small values of axial displacement between the rings during operation, the rubber washer, which has different characteristics from the

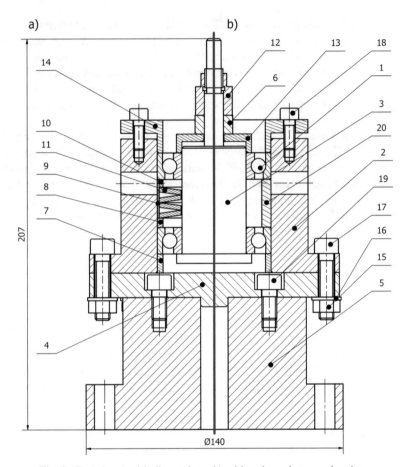

Fig. 2. Test rig: a) with disc springs, b) with a sleeve between bearings.

disc spring package, should not significantly affect the tension force value. Paper [7] presents exemplary solutions for the application of bearing load using rubber elements.

As the rigid preload is the most common way of preloading spindle bearings, it was decided to carry out resistance tests also for this case. The test rig only differed from the above-described version in that the rubber washer was abandoned, so no elastic element was used. In this case, bearing preloading proved to be problematic, as even a small rotation of nut 12 caused high bearing tensioning force. Moreover, it should be mentioned that the value of the force obtained in this way should be interpreted in a slightly different way; namely, this force should be converted according to contact models into the size of relative mutual displacement of bearing rings. The view of the rig with a rigid bearing load is shown in Fig. 2b.

Figure 3 shows a view of the assembled test rig (without cover 14). The rig is screwed to a steel plate; so, it is possible to adjust its rotation axis in relation to the axis of the driving electrospindle.

Fig. 3. Assembled test rig.

3   Measuring Setup and Sample Results

Measurements of motion resistance require the acquisition of data regarding resistance torque, preload, and rotational speed. Below there is a description of measuring setup allowing to record the above-mentioned parameters.

The preload was adjusted every time before the measurement. The preload value, determined through the appropriate tightening of the nut on the shaft, was measured with a Kistler type 9102A piezoelectric sensor. The next device in the measuring setup was the Kistler 5011 charge amplifier. A voltage signal [V] corresponding to the measured force [N] was read directly from the universal meter.

The resistance torque was measured using a Kistler type 9272 piezoelectric sensor, from which the signal was transmitted to a Kistler type 5070 charge amplifier. The next elements of the measuring setup were the analog/digital card and the computer where the signal was recorded in the program developed in the LabView environment. The sampling frequency was of 2048 Hz. Data was processed further in the MS Excel software.

The rotational speed was measured directly by recording the electrical signal from the controller. This signal was recorded in the same program as the resistance torque on the next channel through the analog/digital card.

Figure 4 shows an example of the signal (resistance torque and speed expressed in volts). The measurement was made on the original version of the test rig for preload of 500 N and 3000 rpm. As the above-mentioned figure shows, the interpretation of the results is difficult, if not impossible, therefore each time the moving average method was used to smooth out the course of the torque and speed signal. It is also necessary to correct the offset of the signal obtained each time.

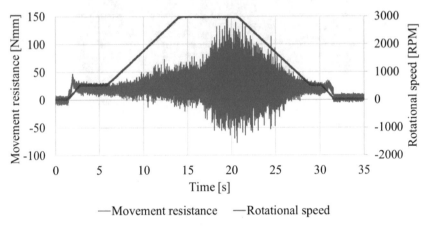

Fig. 4. Sample course of the resistance torque – signal unprocessed.

Figure 5 presents the signal from Fig. 4 after the processing. It is now much easier to interpret the results (torque). From the perspective of the test objective, the most important is the resultant resistance torque of bearing rotating at a fixed speed. Therefore, the mean value of this torque within a certain period of time was calculated when spindle reaches desired rotational speed.

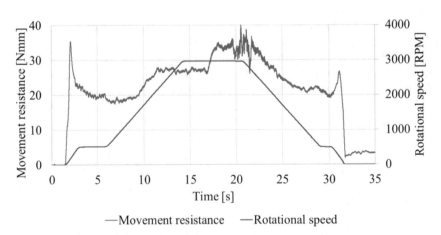

Fig. 5. Sample course of the resistance torque – signal processed.

4   Test Results

The prepared experimental test schedule included the study of the influence of preload and rotational speed on the total resistance torque of bearing motion. The manufacturer states that the permissible B7007-E-T-P4S bearing speed for oil mist lubrication is 34000 rpm [12]. However, due to the capabilities of the rig and safety considerations, the maximum speed was limited to 12000 rpm. The test schedule considered the following speeds: 500, 1000, 3000, 6000, 10000 and 12000 rpm, and preload: 100, 500 and 1000 N. Each measurement was carried out in the shortest possible time to limit the influence of temperature on the results.

Figure 6 shows the results of tests focused on determination of the influence of preload and speed on motion resistance. The tests were carried out on a rig with disc springs. The value of resistance torque refers to a pair of bearings, i.e. a single bearing has twice as low motion resistance. Each measurement was carried out separately and measuring signals were recorded in the period of time in which the shaft starts up, reaches, for example, 6000 rpm, slows down and stops. Then, another analogous measurement was made. Each measurement was repeated at least twice to rule out fatal errors. The results of repeated measurements were, however, similar.

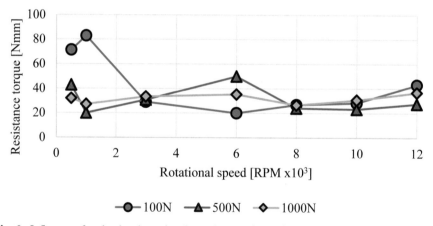

Fig. 6. Influence of preload and rotational speed on motion resistance (test rig with disc springs).

As can be seen in Fig. 6, no clear dependency of motion resistance on preload or speed can be noticed. Based on many of other works [among others 2, 6] it can be assumed that such dependencies exist in reality. The results obtained are close in terms of values regardless of bearing speed and preload, especially at low speeds (500, 1000 and 3000 rpm) and high speeds (8000, 10000 and 12000 rpm).

Figure 7 shows one of the signals recorded at 100 N preload when the bearings are accelerated to 12000 rpm. For low speeds, i.e. 500 to around 2000 RPM, the motion resistance increases and amounts to around 50 Nmm, but after these speeds are exceeded, the motion resistance decreases significantly (at around 6000 RPM it amounts to approximately 20 Nmm), and then the torque increases again until the speed assumed for this test is reached. In terms of qualitative assessment, such course corresponds to the values from Fig. 6. Too many components between the bearings (4 disc springs, 2 spacer rings, and a thin-walled centering sleeve) are a possible cause of a drop in torque value and lack of dependency on speed. Due to number of elements, any positional errors, clearances, etc., between them, makes a large resultant error. Therefore, it is probable that the value of preload changed during operation as a result of the change of relative position of bearing rings in relation to each other. Consequently, it was decided to radically simplify the design of the test rig by replacing all elements between the bearings with a rigid sleeve and adding a rubber washer (see Sect. 2).

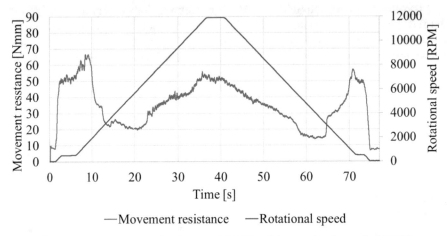

Fig. 7. The resistance torque for preload of 100 N and the rotational speed of 12000 rpm.

Figure 8 shows the results of tests on the influence of preload and speed on the motion resistance while using rubber washer. For these tests, it was decided to reduce the preload slightly in order to increase the durability of the tested bearings. The presented figure shows some clear influence of both the preload and speed on the resistance torque. As stated previously, this could not be observed when it comes to the test results obtained on the test rig with disc springs. This influence is particularly visible for speed higher than 3000 rpm. Apart from that, the resistance torques at preload of 100 N are similar to those obtained on the previous version of the test rig in terms of qualitative assessment. Relatively high resistance at speeds up to 1000 rpm for 100 and 400 N may be caused by hardly recognizable viscous resistance phenomena (the analytical model was developed by Palmgren [11] based on experimental tests). It seems possible that the influence of temperature on the kinematic viscosity of the base oil of the lubricant appeared and affected motion resistance. The temperature in the area of contact between the ball and the raceway depends on speed and load, but also on the duration of the test. Tests for low speeds lasted up to about 20 s, but the higher the speeds, the longer the test time, due to the fixed acceleration and braking ramp. Tests for the maximum speeds lasted over 70 s. Therefore, it is possible that the motion resistance at high speeds is slightly lower due to lower kinematic viscosity. The temperature was not measured during the tests, although even if such measurements were made, the testing time was as minimal as possible.

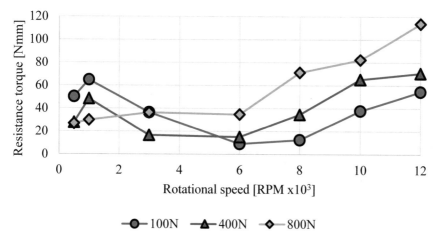

Fig. 8. Influence of preload and rotational speed on resistance torque (the test rig with a rubber washer).

As the spindle bearings are very often preloaded "rigidly", it was also decided to carry out proper tests. This practically did not require any intervention in the design of the test rig, i.e. only the rubber washer was removed. Preloading the bearings for this version of the rig turned out to be a bit difficult, which is explained in Sect. 2 of the paper. Figure 9 shows exemplary influence of speed on the total resistance torque. Also, in this case, each measurement was carried out at least twice.

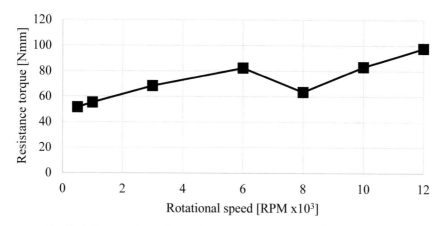

Fig. 9. Influence of speed on resistance on resistance torque (rigid preload).

The presented figure clearly shows that the resistance torque depends on the bearing speed for this type of preload. The original zero speed preload value was 730 N - the value measured with a piezoelectric sensor. As has already been mentioned, in the case of a rigid preload, its value is interpreted slightly differently from the elastic load of bearings. Based on the models of contact between the ball and the raceway, i.e.

contact models [2, 4, 5, 10], the obtained value of force was converted into the mutual relative axial displacement of bearing rings. The value of 730 N corresponds to a load of approximately 12.5 μm. Compared to the elastic load, this load is characterized by a significant increase in the contact forces between the ball and raceway occurring at high speeds. This is caused by high centrifugal forces and the inability of bearing rings to move axially (relative movement along axis). As a result, motion resistance at high and very high speeds has higher values.

Figure 10 shows the motion resistance for the elastic and rigid preload cases, which obtained from the measurements. Up to approximately 6000 rpm, greater motion resistance occurs in the case of rigid preload. In turn, above this speed the resistance was comparable. Due to a similar value of the initial axial preload acting on the bearing, the motion resistance obtained through experiments (especially for the speed 8000 to 12000 rpm) brought comparable values. Higher resistance at rigid load, as mentioned earlier, will manifest itself at even higher rotational speeds. The increase in motion resistance is directly related to the increase in contact forces [9]. In the case of the tested bearings, the speeds for which the measurements were carried out are not particularly high.

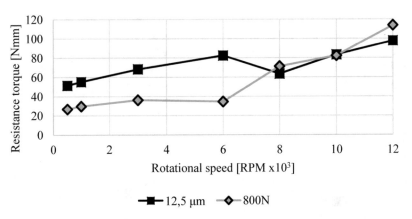

Fig. 10. Comparison of motion resistance for elastic and rigid preload.

5 Summary

The paper presents various concepts of design solutions for a test rig. Not all of the obtained results turned out to be correct, although it seems that the final versions of the test rig gave some satisfying results. If a rubber element in the form of a washer is used, it is possible to adjust the preload relatively easily, however, it should be noted that each elastic preload solution will have lower stiffness of the bearing node comparing to the case of the rigid preload. This is especially true for the preload realized with a susceptible rubber element. However, the measurements were made only at idle motion, so without additional longitudinal loads, which should not affect the operation of bearings and the preload force exerted. The solution used to implement the rigid preload has also proved promising. However, it has been relatively difficult to adjust the preload to an assumed value.

As continuation of the presented research, the problem of heat generation and its influence is currently studied. Heat released inside the electrospindles can significantly affect the machining accuracy. In addition to the amount of heat generated in the motor, bearings are also an important source of heat. The theoretical models that can be applied in this case are quite complex. From other hand, experimental testing of the bearing's resistance torque is a simple way to estimate the amount of heat generated during operation. This justifies intensive work oriented on the practical use of two methods of bearing preloading affecting motion resistance and amount of generated heat, at the same time.

References

1. Abele, E., Altintas, Y., Brecher, C.: Machine tool spindle units. CIRP Ann. Manuf. Technol. **59**(2), 781–802 (2010)
2. Harris, T.A., Kotzalas, M.N.: Rolling Bearing Technology: Advanced Concepts of Bearing Technology, 5th edn. CRC Press, Boca Raton (2007)
3. Hwang, Y.-K., Lee, C.-M.: A review on the preload technology of the rolling bearing for the spindle of machine tools. Int. J. Precis. Eng. Manuf. **11**(3), 491–498 (2010)
4. Jędrzejewski, J., Kwaśny, W.: Modeling of angular contact ball bearings and axial displacements for high-speed spindles. CIRP Ann. Manuf. Technol. **59**(2), 377–382 (2010)
5. Jones, A.B.: A general theory of elastically constrained ball and radial roller bearings under arbitrary load and speed conditions. J. Basic Eng. **82**(2), 309–320 (1960)
6. Kosmol, J.: Determination of motion resistances in high-speed spindle angular bearings. Wydawnictwo Politechniki Śląskiej, Gliwice (2016)
7. Lee, C.-M., Woo, W.-S., Kim, D.-H.: The latest preload technology of machine tool spindles: a review. Int. J. Precis. Eng. Manuf. **18**(11), 1669–1679 (2018)
8. Musiał, J., Styp-Rekowski, M.: Analityczno – eksperymentalny sposób określania współczynników oporów ruchu przy tarciu tocznym. In: Problemy niekonwencjonalnych układów łożyskowych, materiały konferencyjne, pp. 59–65. Łódź (1999)
9. Muszyński, M.: Badania symulacyjne wpływu sprężystych deformacji promieniowych obrotowego pierścienia na siły kontaktowe w łożysku skośnym. Modelowanie Inżynierskie **40**, 63–68 (2019)
10. Noel, D., Ritou, M., Furet, B.: Complete analytical expression of the stiffness matrix of angular contact ball bearings. J. Tribol. **135**(4), 041101-1–041101-8 (2013)
11. Palmgren, A.: Ball and Roller Bearing Engineering, 3rd edn. SKF Industries, Burbank (1959)
12. https://medias.schaeffler.com/medias/en!hp.ec.br.pr/B70..-EB7007-E-T-P4S. Accessed 14 Jan 2020

Example of Using Smoothed Particle Hydrodynamic Method in the Design of Angler Fishing Float

Mariusz Pawlak⬤, Tomasz Machoczek$^{(\boxtimes)}$ ⬤, and Sławomir Duda⬤

Department of Theoretical and Applied Mechanics, Silesian University of Technology, Konarskiego Street 18a, 44-100 Gliwice, Poland
tomasz.machoczek@polsl.pl

Abstract. One of the most popular meshless methods in engineering applications is SPH method, broadly used in the simulation of cutting process and fluid-structure interaction. Examples of this method usage, by a module implemented in the commercial software LS Dyna, are presented by authors. They are compared with experimental tests. In this publication are presented experimental and numerical results from observing the behaviour of the bodies during impact with water. The need to study this phenomenon resulted during the design process of an angler fishing float, which was developed and applied for patent protection by one of the co-authors. For this purpose, a station composed of a transparent water-filled tank attached to a force sensor was created and the numerical model was prepared. Bodies in the form of a SPHere, cube, cylinder, and cone of a constant volume of one litre each were immersed in water at a constant displacement of 100 mm, represented by a quadrant of sinus function at frequencies of 0.5 Hz, 1 Hz, 2 Hz and 3 Hz. The test was performed on a dynamic endurance machine manufactured by MTS 858 Table Top System. Parellely was developed the numerical model with water as SPH elements. Numerical analysis was performed on the mpp (massively parallel processor) version of LS Dyna R11 with the use of 3 nodes and 10 processors per node. Principles of the SPH method implemented in LS Dyna are described in the literature. The purpose of this article was to determine the influence of the body shape immersed in water on the value of the load recorded on the force transducer at the base of the tank. Additionally, estimation of the influence of frequency on loads was to notice. The paper presents the results of experimental studies, which were then compared with the results of numerical simulation.

Keywords: Modelling fishing float · Angler fishing float · SPH method · LS Dyna · Impact

1 Introduction

Floats have different shapes, from bulky and thick to thin and slender depending on the type of fishery: standing water, flowing water, current speed, depth of fishing, and fishing

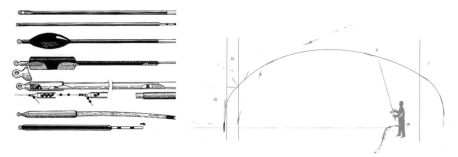

Fig. 1. Selected characteristic shapes of fishing floats [1] and the phase of the correct flight of the float [2]: Waggler type (intended for the so-called "distance method"): a - straight, b - typical with the body, c - with a brass ballast, antenna of material other than the body and with low buoyant signaling tip, d - construction of the club "Cyprinus", e - simple construction with a stem of a cane or a small stick as antenna, f - type "Antenna".

conditions: methods of fishing, type of fishery, depth of fishery and weather conditions. The most frequently used shapes, depending on the application, are shown in Fig. 1:

It is advantageous when the float has a streamlined shape, which provides him with a free and in the distance method, distant flight and minimum viscous resistance in contact with water, in which he fulfills his essential task (signals the moment of the bait being caught by the fish). In addition, the float must allow to perform several flight phases during the stage of casting a fishing set (created by a float suspended on a wire connecting the rod with the hook on which the bait is located). Figure 1 shows the correct flight phases of the wagler type float.

The trajectory as well as the behavior of the float during the flight are shaped by the movement of the rod and the manipulation of the rope on which the float is suspended. The angler gives him four characteristic phases during the casting phase of the set with the float: I - obtained by the rod swing, II - resulting from the free movement obtained on the basis of phase I, III - the float's braking phase, performed to reduce the impact energy of the set against the water surface, primarily for eliminating the potential for entanglement in the fishing set, IV - the phase of free vertical drop (or close perpendicular to the water surface).

The design of the invention (Fig. 2) is based on a base consisting of two parts (1, 2), a polyethylene tube, an antenna (3) with a signaling element in the form of magnetic, color cartridges (4, 6), and seals (5, 9) between the body and the tube and the wobble (8) and the weighters (7) in the float body.

The subject of this paper is an experimental and numerical investigation of the impact of different shape bodies with water. Thanks to this results are possible to select the correct shape body of the fishing floats, used to signal on the surface of the water the moment of catching the bait by the fish. Float geometry is mainly determined by the nature of the signaling moment the bait is consumed by the fish. Each species of fish has its own specific way of collecting food. The cichlid fish, and in particular the big ones do this cautiously, as opposed to predatory and small-sized fish. It is therefore important to shape the float in an optimum manner, taking into account the varied range of fish

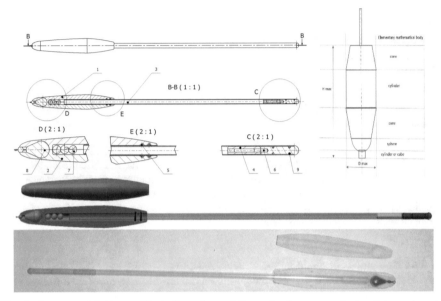

Fig. 2. Structure of fishing tackle dedicated especially to the distance method and the real model of fishing float.

species. The most important parameters characterizing the float, and in particular the floating float, are displacement, aerodynamic, and hydrodynamic resistance. Resistance affects the sensitivity of the signaling of the bite and the tendency to pull the float under the surface of the water by fish, waves or currents, while the aerodynamic resistance determines the range or distance to which we can place the float during ejection using the rod. Hydrodynamic resistance affects the signaling efficiency of the bristles and the depth of immersion and acoustics resulting from the contact of the float with the water mirror at the moment of ejection of the set to the intended distance. It is therefore important to appropriately handle the immersion depth as well as the sound wave, which is usually the unfavorable impact of the float on the surface of the water. Floats have different shapes from thick and short to thin and slender depending on the type of fishery: standing water, flowing water, the speed of flowing water, fishing depth, and fishing conditions: fishing methods, kind of fishery, depth of the fishery and atmospheric conditions. So far, designed and manufactured fishing floats are constructed from a component or components with a small mass and, at the same time, with constant or limited adjustment, a considerable buoyancy, thanks to which it is possible to use the height restriction (float position at the water mirror border) and the drop rate bait in the water. New float constructions provide additional displacement adjustment by changing the weight values directly connected to the float. The materials used for the production of fishing tackle bodies are poplar, willow and lime, polystyrene, sarong, balsa, carbon fiber, natural cork, polyurethane foam, and EVA. The material intended for the fishing float body should be as light as possible, and at the same time characterized by high buoyancy, it should be characterized by low absorbency, oxidation resistance, mechanical strength and ease of processing the material, as most materials used in the production of float bodies are subjected to cavity treatment to give the desired shape.

2   Experimental Hydrodynamic Research

Experimental research was carried out to determine the dynamic load characteristics as a function of immersion of individual bodies while determining the maximum values of forces and the nature of the water mirror stimulation. They were carried out on a test bench based on the MTS 858 Tabletop hydraulic test machine [27] (Fig. 3), that allows for static and dynamic tests with uniaxial compression or tension. The control system of this device consisted of a computer and an MTS FlexTest SE controller, which connected the computer and the test machine to enable the control signal and to process the voltage output into force and displacement in the time domain. The maximum force generated by the hydraulic cylinder of this machine was ±25 kN, and the force sensor allowed it to read to ±15 kN with a resolution of 0.1 N.

Fig. 3. Diagram of a strength machine used for hydrodynamic research.

A tank made of 254 mm (10 in.) internal diameter polycarbonate sleeve was placed on the force transducer, integrated into the machine base and closed from the bottom with a circumferential sealing cover (Fig. 3). A reference grid with a resolution of 5 mm was applied on the cylindrical surface of the tank to determine the size of the wave created as a result of the motion of the indenter. The height of the water column before immersion was 315 mm. The water temperature during the experiment was 19 °C, the atmospheric pressure reduced to the level of the Baltic Sea 101.2 hPa

In addition to the experimental research, a special set of indentations was created in the form of elementary mathematical bodies, i.e. ball, cylinder, cube and cone (Fig. 4). All models were made using an additive technology called FDM (Fused Deposition Modeling), which consists of forming a semiplastic thermoplastic. Solids were made as shell (shell) models with a wall thickness of 2 mm and an internal volume of 1 dm^3. Water for better observation of wave propagation was stained with tea.

Fig. 4. List of indenters with their geometric parameters and tests no.

3 Numerical Model

During the process of modelling, various methods were considered, from Euler method, through ALE method, to SPH method [19, 21–26], which looked the most promising due to both Lagrange and Eulerian character. SPH method is used widely in engineering applications, from water wave modelling, tsunami phenomena, material cutting, or fracture to explosions [1–5, 20]. However must be admitted that grid-based CFD methods may allow better accuracy. Aspects of buoyancy and fluid structure interaction by SPH elements are also considered in the literature [6–8]. Constantly new ideas and improvements are implemented in this meshless method [9–11, 13, 14], which has an advantage comparing to the standard finite element method, that element does not distort. Water material properties are well known and were taken from publications [17, 18].

Simulation of body impact with water is quite popular [12, 16], quite new is the purpose of using it in the design process of angler fishing floats.

To prepare the numerical model precisely, all components used in experiments were measured and checked their mass to receive adequate numerical equivalence (Fig. 5).

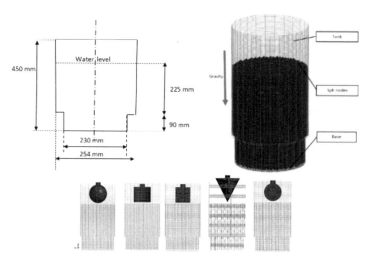

Fig. 5. The geometry of a tank filled with water.

According to calculations, the weight of the empty tank was 65 N, the volume of water was 0.01 m³ and when multiplied by density 998.2 kg/m³ a mass of 15.11 kg is received. In earth gravity equal to 9.806 the weight of water in the tank was 148.2 N. Summarizing this load, the total weight of tank and water should be about 213.2 N.

The numerical model was created in commercial software LS Dyna [1, 15]. where Smoothed particle hydrodynamic method is implemented. It is a Lagrangian collocative explicit method, and in each iteration are solved the conservation of mass, conservation of momentum, and conservation of energy. Accuracy is low compare to finite element or EFG method. Particle model is characterized by a mass m, distance d between particles, and smoothing length h. Smoothing length varies in time and in space.

Parameters d and h are used in calculation of smoothing function (Fig. 6), which should be centrally peaked. Common smoothing kernel is cubic B-spline defined as follows:

$$W(d, h) = \frac{1}{h^{\alpha}(x, y)} \theta\left(\frac{d}{h(x, y)}\right) \tag{1}$$

$$\theta(d) = Cx \begin{cases} 1 - \frac{3}{2}d^2 + \frac{3}{4}d^3 \, for \, /d/ \leq 1 \\ \frac{1}{4}(2 - d)^3 \, for \, 1 < /d/ \leq 2 \\ 0 \, for \, /d/ > 2 \end{cases} \tag{2}$$

where C is a constant of normalization dependent on the space dimension, α is the number of space dimensions.

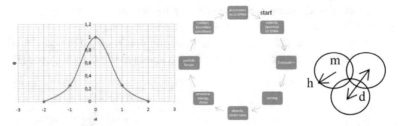

Fig. 6. Smoothing function, cycle loop of SPH method in LS Dyna, d and h parameters.

Shape of smoothing function for $C = 1$

SPH method is based on the quadrature formula for moving particles, where xi(t) is the location of particle i, which moves along the velocity field v.

The particle approximation of a function can be defined as:

$$\Pi^h u(x_i) = \sum_j \frac{m_j}{\rho_j} u(x_j) w(x_i - x_j, h) \tag{3}$$

particle approximation of the gradient

$$\Pi^h \nabla u(x_i) = \sum_j \frac{m_j}{\rho_j} u(x_j) A_{ij} - u(x_i) A_{ji} \tag{4}$$

The advantage of using LS Dyna is the ability of interaction of SPH elements with parts modelled by finite element as shells and solids. The described problem could be solved also with ALE method, but it would be more time consuming and impossible to receive the contact loads at the bottom of the tank. During the process of discretization, the water was divided to 29583 SPH nodes, where the node spacing was 5 mm and mass of each SPH node was assumed 5.11e−04 kg.

Because in experimental tests the force sensor has been reset to zero at the weight of the tank and water, the simulation process had to be separated into two parts; first called dynamic relaxation and the second, where a blast of solids with water occurred. During dynamic relaxation on water was working only gravity, and after 5 s the force in contact between the tank and the base (in the real object there is placed force sensor) was stabilized. The results presented below are in agreement with the real weight of the tank with water (Fig. 7, Fig. 8, Fig. 9 and Fig. 10).

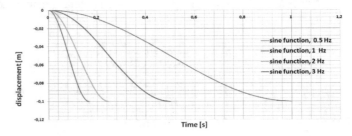

Fig. 7. Displacement of solids along the z axis as a time function at frequency 0.5 Hz, 1 Hz, 2 Hz and 3 Hz.

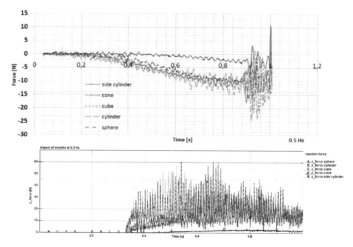

Fig. 8. Comparing results from experiment (top) and simulation (bottom) at frequency 0.5 Hz.

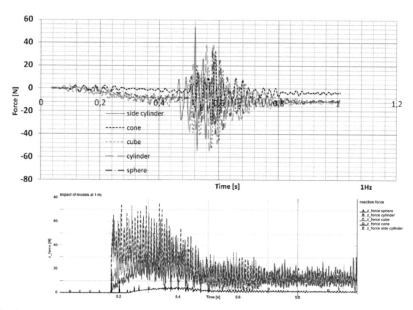

Fig. 9. Comparing results from experiment (top) and simulation (bottom) at frequency 1 Hz.

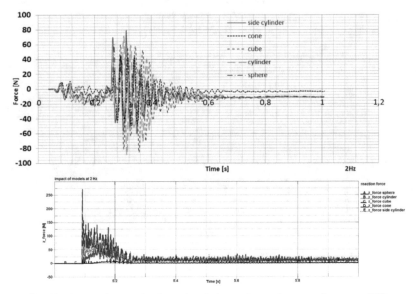

Fig. 10. Comparing results from simulation and experiment at frequency 2 Hz.

4 Comparing Results from Measurements with Numerical Simulation

When taking into account the restrictions mentioned before, from figures presented below can be noticed, that the maximum loads from numerical results are in good agreement with those from experiments.

5 Conclusion

On the basis of the experimental studies, differences were observed in the trajectory of the bodies during impact with water, which results from the presence of air bubbles located at the ends of the hydraulic cylinder of the endurance machine. In this experiment, it was noticed standard delays caused by the transmission and processing of control signals. In the response charts of the immersion pattern of the individual bodies, also a wave excitation on the surface of the water was observed. Results from the numerical model were compared with results from experiments. Immersion tests allow to compare the results of displacement loads and thus to form the basis for the optimization of the shape and geometry of the target fly fishing in the future.

It has been noticed that the experiments conducted five times are characterized by slight differences in the courses of values obtained. The main discrepancies result from the specifics of the test, during which there is a rapid increase in the load, which is associated with the operation of the endurance machine being carried out, and the current setting of control parameters of its regulator.

In addition, it can be observed that with the increase of the forcing speed, the value of the uncertainty error increases, which is the reason for the dynamically changing PID regulator settings as a result of the increase of the pressure force of the analyzed body to the liquid filling the tank.

The noise observed in the graphs of loads as a function of displacement and time is caused by the presence of secondary effects, i.e., a wave reflected from the walls of the vessel, as well as the phenomenon of cavitation formed on the edges of cubic cubes, and the cylinder in the two analyzed locations in places of pressure drop. The secondary effects mentioned in the final analysis do not play a significant role in relation to the emerging cavitation. Cavitation in the case of a fishing float is disadvantageous due to the increased hydrodynamic resistance that can cause scaring of calm fish.

The experimental tests were carried out and the results obtained on their basis can be used to verify and validate the numerical models of the float for the needs of optimizing its shape and geometry.

Acknowledgments. Calculations were carried out using the computer cluster Ziemowit (http://www.ziemowit.hpc.polsl.pl) funded by the Silesian BIO-FARMA project No. POIG.02.01.00-00-166/08 in the Computational Biology and Bioinformatics Laboratory of the Biotechnology Centre in the Silesian University of Technology.

Conflict of Interest. "The authors declare that they have no conflict of interest".

References

1. Lacome, J.L.: Smoothed Particle Hydrodynamics –Part 1 and Part 2, theory manual for SPH in LS Dyna
2. Randles, P.W., Libersky, L.D.: Smoothed particle hydrodynamics: some recent improvements and applications. Comput. Meth. Appl. Mech. Engrg. **139**, 375–408 (1996)
3. Olleak, A.A., El-Hofv, H.A.: SPH Modelling of Cutting Forces while Turning of TI6Al4V Alloy
4. Olejnik, M., Szewc, K., Pozorski, J.: SPH with dynamical smoothing length adjustment based on the local flow kinematics. J. Comput. Phys. **348**, 23–44 (2017)
5. Liu, M.B., Liu, G.R., Zhong, Z., Lam, K.Y.: Computer simulation of high explosive explosion using smoothed particle hydrodynamics methodology. Comput. Fluids **32**, 305–322 (2003)
6. Marrone, S., Bouscasse, B., Colagrossi, A., Antuono, M.: Study of ship wave breaking patterns using 3D parallel SPH simulations. Comput. Fluids **69**, 54–66 (2012)
7. Wu, T.-R., Chu, C.-R., Huang, C.-J., Wang, C.-Y., Chien, S.-Y., Chen, M.-Z.: A two-way coupled simulation of moving solids in free-surface flows. Comput. Fluids **100**, 347–355 (2014)
8. Losasso, F., Talton, J.O., Kwatra, N., Fedkiw, R.: Two-way coupled SPH and particle level set fluid simulation. IEEE Trans. Vis. Comput. Graph. **14**(4), 804 (2008)
9. Liu, M.B., Liu, G.R.: Restoring particle consistency in smoothed particle hydrodynamics. Appl. Numer. Math. **56**, 19–36 (2006)
10. Fang, J., Parriaux, A., Rentschler, M., Ancey, C.: Improved SPH methods for simulating free surface flows of viscous fluids. Appl. Numer. Math. **59**, 251–271 (2009)
11. Amicarelli, A., Albano, R., Mirauda, D., Agate, G., Sole, A., Guandalini, R.: A Smoothed Particle Hydrodynamics model for 3D solid body transport in free surface flows. Comput. Fluids **116**, 205–228 (2015)

12. Shadloo, M.S., Oger, G., Le Touze, D.: Smoothed particle hydrodynamics method for fluid flows, toward industrial applications: motivations, current state, and challenges. Comput. Fluids **136**, 11–34 (2016)
13. Bouscasse, B., Colagrossi, A., Marrone, S., Souto-Iglesias, A.: SPH modelling of viscous flow past a circular cylinder interacting with a free surface. Comput. Fluids **146**, 190–212 (2017)
14. Molteni, D., Vitanza, E., Battaglia, O.R.: Smoothed particles hydrodynamics numerical simulations of droplets walking on viscous vibrating liquid. Comput. Fluids **156**, 449–455 (2017)
15. Xu, J., Wang, J.: Interaction methods for the SPH parts (multiphase flows, solid bodies) in LS-DYNA. In: 13th International LS-DYNA Users Conference
16. Younghu, W., Dongwei, S., Fujii, Y., Takita, A., Araki, R., Wei, H.: Experimental and numerical study of water impact investigations for aircraft crashworthiness application analysis. In: 11th World Congress on Structural Multidisciplinary Optimisation, Sydney Australia, 07–12 June 2015
17. Keegan, M.H.: Doctoral thesis, Wind Turbine Blade Leading Edge Erosion: An investigation of rain droplet and hailstone impact induced damage mechanisms, University of Strathclyde, Glasgow, Scotland (2014)
18. LS-DYNA Aerospace Working Group, Modeling Guideline Document, Version 16-2, 30 December 2016
19. Griebel, M., Schweitzer, M.A.: Meshfree methods for partial differential equations. Lecture Notes in Computational Science and Engineering, vol. 26. Springer (2015)
20. Liu, S.J.: Fracture of thin pipes with SPH shell formulation. Int. J. Comput. Methods **8**(3), 369–395 (2011)
21. Gu, Y.T.: Meshfree methods and their comparisons. Int. J. Comput. Methods **2**(4), 477–515 (2005)
22. Omidvar, P., Norouzi, H.: Smoothed Particle Hydrodynamics for water wave propagation in a channel. Int. J. Modern Phys. C **26**(8) (2015)
23. Wang, J.: Penetration simulation for an open caisson using mesh-free SPH method. In: Computer Methods and Recent Advances in Geomechanics. Taylor & Francis Group, London (2015)
24. Huang, Y., Dai, Z., Zhang, W.: Geo-disaster Modeling and Analysis: An SPH-Based Approach. Springer, Heidelberg (2014)
25. Liu, G.R., Liu, M.B.: Smoothed Particle Hydrodynamics: A Meshfree Particle Method. World Scientific, New Jersey (2003)
26. Violeau, D.: Fluid Mechanics and the SPH Method, Theory and Applications. Oxford University Press, Oxford (2012)
27. https://www.upc.edu/sct/documents_equipament/d_77_id-412.pdf. Accessed 08 May 2020

Analysis of the Impact of Dynamic Loads on Transmission Shafts of a Civil Aircraft

Michal Ptak and Jerzy Czmochowski[✉]

Wroclaw University of Science and Technology, 50371 Wroclaw, Poland
{michal.ptak,jerzy.czmochowski}@pwr.edu.pl

Abstract. The main attention of this article is focused on dynamic loads affecting transmission shafts in the slat and flap transmission systems as well as undesirable effects resulting from these interactions. The article refers to the RTCA aviation standard DO-160G (Environmental Conditions and Test Procedures for Airborne Equipment) regulating the requirements of individual aircraft also from the point of view of vibration in individual zones of the aircraft. The article will focus on two basic load cases: high-level, short duration transient vibration that occurs during an engine fan blade loss and the random vibrations case on an aircraft related to take-off, flight and landing described by the function of power spectral density.

Keywords: Harmonic analysis · Random vibration analysis · Finite element method · Transmission shafts of aircraft

1 Introduction

The article will analyze the impact of dynamic loads on the transmission shafts of a civil aircraft. Transmission shafts are a torque transmitter from a power drive unit to the panels of slats and flaps. The example architecture of both systems (see Fig. 1) shows the arrangement of transmission shafts connected to individual angular and planetary gears of the slat and flap systems. The significance of this system is important for obtaining the aircraft lift force during take-off and landing. These systems allow to increase the lift force of the aircraft at low speeds, making it easier to take off the aircraft and maintaining stability when approaching the landing. In addition, these systems allow the braking distance to be reduced, due to the aerodynamic drag caused during landing, which translates into less effort on the landing gear and brakes of the aircraft and a reduction in the braking distance.

The slat system is located in the front of the wing, while the flap system is located in the rear of the wing (see Fig. 1). Both of these systems are responsible for creating appropriate configurations of the wing profile, adapting it to individual flight stages so as to effectively use the aerodynamic phenomena occurring on the surface of the aircraft wing.

This paper present first stage of research focused on vibration damage assessment under deterministic vibration loading in transmission shaft. First stage of research is preparing numerical environment and their correlation with test results. This allow to

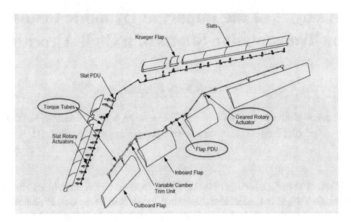

Fig. 1. Transmission shafts architecture in a slot and flap system [1].

assess the impact dynamic load on shaft structure and stress level in aircraft shafts. Correlated numerical environment will be than used for analysis impact of random load on transmission shaft structure and next stage of research which is developing program for vibration damage assessment using Python programing language. This approach is known as model-based method introduced in [2] and will be used and developed in further research.

Main point of research introduced in this paper is assessment the damage under stochastic and deterministic loading in transmission shaft structure. Issue how to assess the damage under deterministic and random loading have been introduced in papers [3] and [4] – this paper present several approach for vibration damage calculation and benchmark on track cabin, which is not relative to the transmission shaft. The conducting research will be focused on verification, which methods will be most accurate and potentially can be modified to the transmission shaft structure and can give results close to the test. Research will be focused on possible developing new method for vibration damage assessment and study new simplified and efficient method for example bands method, introduced in paper [5].

Transmission shafts are welded structures – flanges is welded to the shaft beam, due to that influence from welding residuals stress and their impact on fatigue life which can be not negligible as introduce in [6] will be also taken into consideration in future research.

2   Analysis of Dynamic Load

The aircraft is exposed to dynamic loads, which are a consequence of various phenomena during take-off, flight and landing. During a flight, under normal conditions, dynamic loads resulting from turbulence and vibrations caused by turbojet engines. More attention will be paid to these loads in Sect. 2.1. Turbojets are exposed to various dynamic effect e.g. engine blade loss, which causes the wing system to fall into harmonic vibrations with high acceleration amplitude in the low frequency bandwidth and this load case is one of the most important for transmission shafts. The international aviation standard [7] defines seven vibration zones, which are characterized by individual requirements due to the level of dynamic loads of both harmonic load during engine failure and loads generated by normal flight conditions. Vibration zones related to the location of transmission shafts will be analyzed.

2.1 Random Vibration Case

Random vibrations are normally described in the form of the Power Spectra Density (PSD) function, separately for each vibration zone of the aircraft in the form of appropriate curves (see Fig. 2). In the case of transmission shafts, the C/C1 curves are used for locations in the aircraft fuselage, while for location on the wing of the aircraft the more restrictive curves E/E1 are used, this is due to the greater susceptibility of the aircraft wing to vibrations of different source. The component exposure time to individual load profiles is 3 h for each orthogonal direction.

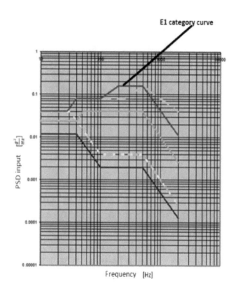

Fig. 2. Spectral density of signal strength for individual standard curves.

The more critical E/E1 curves indicates the maximum spectral density of the signal power in the 200–500 Hz frequency bandwidth, equal to $0.16\,g^2/Hz$. Numerical analyzes on various components of the aircraft, in different vibration zones have shown that the components are most damaged in this bandwidth in case of random vibration load case. Transmission shafts rarely achieve maximum gain in the 200–500 Hz bandwidth, which means that they are less exposed to the most energized bandwidth. As a rule, these are components showing high axial and transverse rigidity, e.g. bodies of various types of mechanisms. In the case of transmission shafts, numerous numerical analyzes prove that the effort is secondary compared to the case of a turbojet turbine rotor detachment and this case will be discussed in the another publication.

2.2 High Level Short Duration (HLSD) Vibration Case (High Constant Amplitude Harmonic Excitation)

Harmonic and high amplitude forcing caused by a turbocharger rotor rupture is extremely dangerous for transmission shafts. Transmission shafts, which are a slim design with relatively low transverse stiffness, due to their length have a high flexural susceptibility. This causes that during resonance in the low frequency bandwidth, the amplitudes of the shaft center point exceed a few or even several centimeters from the neutral axis in the nominal position. An excited shaft can be a threat to other strategic aircraft systems located in the wing and in the fuselage, high deformation may cause collision of the shaft with metal or composite structural elements may initiate cracks or delamination of the material. In addition, stress in the shaft resulting from high lateral deformation requires attention during analyzes at the design stage.

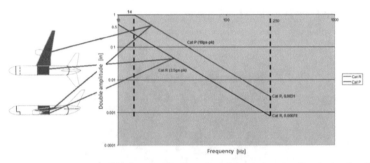

Fig. 3. Load curves for considered transmission shaft locations, amplitude in the frequency domain.

In this publication, the main focus will be on HLSD type of load. The sample transmission shaft located on the aircraft wing will be analyzed (P curve) as a more critical case than the location in the fuselage (R curve) see Fig. 3.

Harmonic forcing has a sinusoidal shape with constant amplitude. Sweep rate for harmonic excitation is defined in a linear form as 0.167 Hz/s. This parameter will be taken into account only at the stage of quantitative vibration estimation of the fatigue of the shaft material.

The equation of displacement in steady motion can be written as Eq. (1) according to [8]:

$$x = Xsin(\omega t + \phi) \tag{1}$$

X - amplitude of displacement, ω - circular frequency.

The speed in harmonic motion is a derivative of Eq. (1):

$$\dot{x} = \frac{dx}{dt} = \omega Xcos(\omega t + \phi) \tag{2}$$

ωX - amplitude of velocity.

Acceleration in harmonic motion is a derivative of Eq. (2):

$$\ddot{x} = \frac{d^2x}{dt^2} = -\omega^2 X sin(\omega t + \phi) \tag{3}$$

$-\omega^2 X$ – amplitude of acceleration.

The definition of the constant acceleration amplitude as a function of frequency f for the purpose of numerical calculations is implemented by the following Eq. (4):

$$-\omega^2 X = X \cdot (-2\pi \cdot f)^2 = const \tag{4}$$

The characteristic points of the P curve are defined by the values of the frequency f_p and the corresponding values of the vibration amplitude $X_{Amplitude\ of\ displacement\ P}$

$$f_P = \begin{pmatrix} 10 \\ 14 \\ 250 \end{pmatrix} Hz, X_{Amplitude\ of\ displacement\ P} = \begin{pmatrix} 12.7 \\ 12.7 \\ 0.04 \end{pmatrix} mm \tag{5}$$

Acceleration amplitudes determined according to Eq. (4) for characteristic points will be:

$$Acc_{Amplitude\ of\ acceleration\ P} = \begin{pmatrix} 5.01 \cdot 10^4 \\ 9.81 \cdot 10^4 \\ 9.81 \cdot 10^4 \end{pmatrix} mm/s^2 \tag{6}$$

In gravitational acceleration units it will be:

$$Acc\_g_{Amplitude\ of\ acceleration\ P} = \begin{pmatrix} 5 \\ 10 \\ 10 \end{pmatrix} g \tag{7}$$

Load definition from equitation (7) will be implemented in numerical model.

3   Application of the Steady State Dynamics Theory for Problem Analysis

In this publication the Steady State Vibration Theory will be used as a basis for further numerical calculation. The basis for describing the dynamic behavior of the system is the Lagrange equation of motion in the form [9]:

$$m\ddot{x}(t) + b\dot{x}(t) + kx(t) = F_0 \cdot sin(\omega t) \tag{8}$$

m - mass, b - viscous damping factor, k - linear stiffness, ω - frequency of excitation force.

Which can be demonstrated as a single degree of freedom system (see Fig. 4)
Equation (8) can be written in the equivalent form:

$$m\ddot{x}(t) + b\dot{x}(t) + kx(t) = ma_0 \cdot sin(\omega t) \tag{9}$$

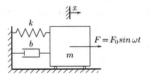

Fig. 4. Harmonic motion of a system with one degree of freedom.

The solution of the differential equation (9) can be presented in the form of the following equation:

$$x(t) = X \cdot \sin(\omega t - \Phi) + X_1 \cdot e^{-\zeta \omega_n t} \cdot \sin\left(\sqrt{1 - \zeta^2} \omega_n t + \Phi_i\right) \qquad (10)$$

And its graphical representation [8] (see Fig. 5):

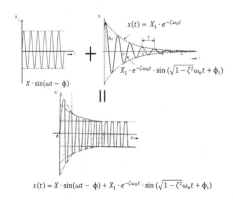

Fig. 5. Graphic representation of steady state and transient superposition based on Eq. (10).

$$X = \frac{F_0/k}{\sqrt{\left(1 - \frac{m\omega^2}{k}\right)^2 + \left(\frac{c\omega}{k}\right)^2}} \qquad (11)$$

ω – force frequency, ω_n – natural frequency of the system, $\zeta = c/c_c$ – damping factor, $c_c = 2m\omega_n$ – critical damping factor, Φ – phase shift of the signal,
$X \cdot \sin(\omega t - \Phi)$ – special solution of differential equation - steady state equation
$X_1 \cdot e^{\zeta \omega_n t} \cdot \sin(\omega t + \Phi_i)$ – homogeneous solution of differential equation - transient oscillations.

The standard loads [7] determine the requirements as stationary excitation with constant amplitude, which implies the possibility of omitting the effects associated with non-stationary vibrations during the implementation of harmonic excitation. This, in turn, leads to the possibility of linearizing the system and the possibility of performing calculations in the frequency domain, which are characterized by a much lower cost than non-constant calculations with explicit integration of motion equations from [10]. In addition, non-stationary effects are expected during the initiation of harmonic motion at frequencies up to 14 Hz, which is no critical, due to the fact that there is no resonance around this frequency, which is one of the design criteria at an early stage of shaft design.

After considering the stationary part of displacement, the gain factor can be defined as:

$$\frac{X}{X_0} = \frac{1}{\sqrt{(1 - (\frac{\omega}{\omega_n})^2)^2 + (2\zeta \frac{\omega}{\omega_n})^2}} \tag{12}$$

X_0 – static amplitude – for zero frequency.

Equation (12) will be used to estimate the damping using the Half-Power Bandwidth method introduced in [11] as applicable to systems with low damping such as transmission shafts. The bandwidth according to this method is measured as 0.707 maximum amplitude. The scheme of mentioned method introduced in [12] is presented below (see Fig. 6).

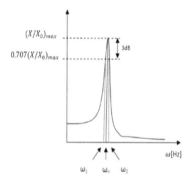

Fig. 6. Half Power Bandwidth method scheme, [12].

For the resonance frequency: $\frac{\omega}{\omega_n} = 1$, $\frac{X}{X_0} = \frac{1}{2\zeta}$, based on equitation (12). Then the frequency ω_1 and ω_2 (see Fig. 6) is searched for such that $\frac{X}{X_0} = 0.707\frac{1}{2\zeta}$, as introduced in [12] after substitution into Eq. (12), the equation in the form of a fourth degree polynomial (13) is obtained:

$$(\frac{\omega}{\omega_n})^4 - 2\left(1 - 2\zeta^2\right)(\frac{\omega}{\omega_n})^2 + \left(1 - 8\zeta^2\right) = 0 \tag{13}$$

It's solution is the quadratic equation:

$$(\frac{\omega}{\omega_n})^2 = \left(1 - 2\zeta^2\right)(\frac{\omega}{\omega_n})^2 \pm 2\zeta\sqrt{(1 + \zeta^2}) \tag{14}$$

Assuming small values of damping for the structure under consideration, we obtain:

$$(\frac{\omega}{\omega_n})^2 = 1 \pm 2\zeta \tag{15}$$

For two roots of function (15) ω_1, ω_2 we obtain two solution: $(\frac{\omega_1}{\omega_n})^2 = 1 - 2\zeta$, $(\frac{\omega_2}{\omega_n})^2 = 1 + 2\zeta$.

After deducting them by sides, the Eq. (16) is obtained:

$$(\frac{\omega_2}{\omega_n})^2 - (\frac{\omega_1}{\omega_n})^2 = 4\zeta \tag{16}$$

Which, after transformations, leads to Eq. (17):

$$\frac{\omega_2 - \omega_1}{2\omega_n} = \zeta \tag{17}$$

Equation (17) will now be used to estimate the initial damping value based on the gain characteristic in the frequency domain (obtained during the harmonic excitation of the real object) for individual resonance frequencies before starting the numerical procedure.

4 Correlation Numerical Environment with Test Data

Numerical model is constructed to reflect the test procedure defined by the aviation standard [7]. The initial testing phase is searching for resonance in the $0 \div 2000$ Hz bandwidth with a signal of constant acceleration amplitude equal to 0.5g. To reflect the course of the test in the numerical model, a modal analysis was performed to estimate the resonance frequencies in the tested bandwidth. The results of modal analysis are input data for harmonic analysis with constant acceleration of 0.5g using modal superposition procedures introduced in [10] and [13].

The discrete model is a combination of a solid model, discretized with hexagonal elements with a linear shape function and a shell model discretized with shell elements, four nodes with five integration points on thickness.

Solid and shell elements have been bonded in the ABAQUS numerical environment [13].

The damping value at individual resonance frequencies was initially estimated according to the theory in paragraph 3, and then it was modified so that the excitation levels obtained from the numerical procedure were close to the signal gain on a real object. The waveform of 0.5g signal amplification in the frequency domain for the transverse direction (Y) is presented below (see Fig. 7).

Due to the symmetry of the analyzed transmission shaft, the transverse direction perpendicular to the tested one will not be quoted in this publication because of the negligible differences of the tested parameters.

Similar steps were used when correlating the numerical environment in an axial direction. The characteristics of the system response to acceleration in the axial direction during forcing in this direction are presented below (see Fig. 8).

The first axial resonance will occur at a frequency of about 550 Hz, which means that this load case is not as significant when analyzing the 10g forcing case for the HLSD case, which is applied in the frequency range of 0–250 Hz. Nevertheless, it will be relevant to the case of random vibration load case, where for this frequency the function of signal power spectral density reaches its maximum value.

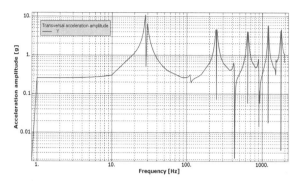

Fig. 7. Characteristics of shaft center point acceleration in the frequency domain in the Y axis, excitation in the Y direction.

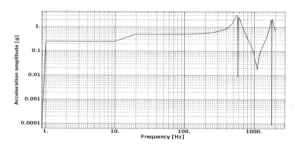

Fig. 8. The acceleration characteristics of the shaft center point in the frequency domain in the Z axis, forcing in the Z direction.

Numerical model of considered shaft is correlated to the test performed on shaft, acceleration amplitude magnification during the most severe resonance for transverse direction as approximately 22 (input 0.5g output 11g) and for axial direction approximately 6 (input 0.5g output 6g). Additionally all other resonances occurred during the test are replicated in numerical model and their amplification is close to the test amplification.

5   Finite Element Analysis of Transmission Shaft for Critical Load Case

This paragraph presents the results of analyzes for 10g acceleration assuming that the tested system is linear even for such large harmonic excitations. This does not mean, however, that future research, analyzes and publications will not focus on the analysis of system non-linearity during critical excitations defined by the load case under examination.

The maximum peak acceleration of the reference point placed in the plane of the shaft symmetry during low-frequency resonance - about 29 Hz is 216.6 g, (see Fig. 9) it is a value that suggests that nonlinear effects (e.g. geometry nonlinearity) might not be ignored.

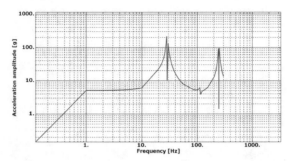

Fig. 9. Characteristics of shaft center point acceleration in the frequency domain in the Y axis, excitation in the Y direction.

The lateral displacement corresponding to this acceleration is about 70 mm, (see Fig. 10).

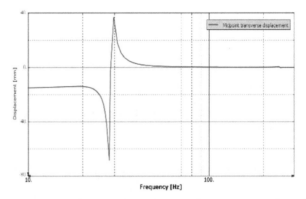

Fig. 10. Total displacement of the reference point located on the outer surface of the shaft in the plane of symmetry.

The reduced Huber-Mises-Hencky stress for the first resonance frequency is about 680 MPa (see Fig. 11 and Fig. 12) which is a relatively high result with the tensile strength of the shaft material around 1000 MPa. This is the stress result for the outer surface of the shaft, in addition, the following figure presents the results of stress in the frequency domain for the inner surface of the shaft (see Fig. 11).

Due to significant shaft displacements, further analyzes and experimental tests will be undertaken to study the non-linearity of structural behavior under critical harmonic excitations. Research will focus on testing the level of damping and its dependence on the excitation force.

Similar numerical calculations were carried out for the same value of the constant amplitude for harmonic excitation in the axial direction. The frequency characteristic in axial direction is recorded below (see Fig. 13), the maximum acceleration in the tested bandwidth was about 12g. This is due to the fact that axial resonance occurs at high frequencies in the range of 500 ÷ 600 Hz and excitation in the axial direction is not

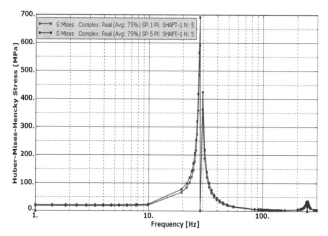

Fig. 11. Huber-Mises-Hencky reduced stress in the critical node, results for integration point 1 - inside the shaft (SP: 1) and point 5 - outside the shaft (SP: 5) in the frequency domain.

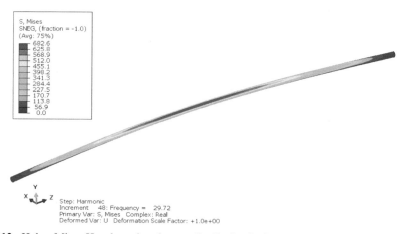

Fig. 12. Huber-Mises-Hencky reduced stress distribution in the most stressed central part of the transmission shaft.

critical for this case. Nevertheless, it will be more important for the case of random vibrations, because the resonance in this bandwidth is superimposed with the maximum signal strength (see Fig. 2).

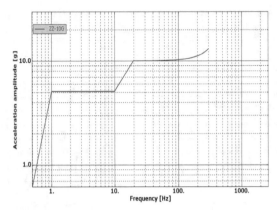

Fig. 13. The acceleration characteristics of the shaft center point in the frequency domain in the Z axis, forcing in the Z direction.

6   Discussion of Results, Conclusions and Sketching the Direction of Further Research and Analysis

The harmonic analysis of the transmission shaft assembly, which is an integral part of the civil aircraft slat and flap systems, showed a significant level of acceleration amplitude at low frequency resonance. Analyzes have shown that the first shaft resonance occurs at low frequency - about 29 Hz, resonance in such low frequency case high deflection, which implies high stress at outer diameter of shaft. The high level of amplitude causes high stress value at the outer surface of shaft, which can affect fatigue life during the occurrence of a critical load case. In addition, shaft deflection is high (single amplitude approximately 70 mm) and it will be necessary to undertake further analyzes in the field of estimating the non-linear behavior of the system. Another resonances in higher frequency are no critical for HLSD load case, because are above the maximum frequency when the HLSD acceleration is applied, however more severe for random vibration case – the maximum power spectra density of signal is for bandwidth from 200 through 700 Hz. For this reason, future research and analysis will also focus on the quantitative estimation of structural damage using Python language of programing based on the results of harmonic analysis, random vibrations and the test procedure defined by the international aviation standard [7]. The Python procedures will be developed and benchmarked using test results for representative group of shafts.

References

1. Australian Government: Tyre failure and flap asymmetry event involving Boeing 787. VH-VKA. Australian Transport Safety Bureau, Singapore (2017)
2. Fan, W., Qiao, P.: Vibration–based damage identification methods: a review and comparative study. Struc. Health Monitor. **10**, 83–111 (2011)
3. Bishop, N., Murthy, P., Sweitzer, K.: Advances relating to fatigue calculation for combined random and deterministic loads. In: 13th International ASTM/ESIS Symposium on Fatigue and Fracture Mechanics (39th National Symposium on Fatigue and Fracture Mechanics), Jacksonville, FL, 13–15 November 2013

4. Bishop, N., Sweitzer, K., Schlesinger, D., Woodward, A.: Fatigue calculation for multi input random and deterministic loads in the frequency domain. In: UK NAFEMS Conference, Oxford IK, Accelerating the Future of CAE, 10–11 June 2014
5. Braccesi, C., Cianetti, F., Tomassini, L.: Random fatigue. A new frequency domain criterion for the damage evaluation of mechanical components. Int. J. Fatigue **70**, 417–427 (2015)
6. Mi, C., Liu, J., Xiao, X., Li, W., Sun, X., Ming, X., Tao, C.: Random vibration fatigue life assessment and optimalization of train buffer beam considering welding residual stress. J. Mech. Sci. Technol. **34**(3), 1071–1080 (2020)
7. RTCA, Inc.: Environmental Conditions and Test Procedures for Airborne Equipment. "RTCA/DO-160G". RTCA Incorporation, Washington DC (2020)
8. Benson, H.T.: Principles of Vibration. Oxford University Press, New York (1996)
9. Thomson, W.T.: Vibration Theory and Applications. Prentice-Hall, Inc., Englewood Cliffs (1965)
10. Zienkiewicz, O.C.: Finite Element Method. McGraw-Hill, Dallas (1977)
11. Kelly, G.S.: Fundamentals of Mechanical Vibrations. McGraw-Hill Higher Education, Boston (2000)
12. Dimargonas, A.: Vibrations for Engineers. Prentice Hall, Inc. Upper Saddle River (1996)
13. ABAQUS User Manual V6.10-1. Dassault Systems (2010)

Numerical Analysis of Fragmentation and Bullet Resistance of Various Steel Thicknesses and Composite Plates

Kamil Sadowski, Edyta Krzystała$^{(\boxtimes)}$ ⓘ, and Sebastian Sławski ⓘ

Faculty of Mechanical Engineering, Silesian University of Technology, Konarskiego street 18A, Gliwice, Poland
Edyta.Krzystala@polsl.pl

Abstract. The paper presents the results of finite element research concerning the fragmentation and bullet resistance of steel plates with various thicknesses as well as epoxy/aramid composite plate. Due to the large number of armed conflicts in various parts of the world, this issue is extremely important due to the safety of infantry soldiers in the areas of military operations. The work presents the results of preliminary numerical tests carried out in commercially available software using the finite element method. The fragment and bullet resistance of three steel plates with various thicknesses at different impact speeds is compared. The material model that was used for steel plates considers the elastically plastic behaviour of the material. Standard striking elements were used for both fracture and bulletproofing. In both cases, Johnson Cook's material model was used, which allows observation of the mushrooming process during impact. The article compares the damage caused by the impact of the bullet core and 9×19 mm bullet and then specifies the conditions for which steel plates meet the requirements of fragmentation and bullet resistance in accordance with PN-V 87001:12011. What is more, epoxy/aramid composite was analyzed as well.

Keywords: Finite element method · Fragmentation-resistant analysis · Numerical simulation · Ballistic protection

1 Introduction

Protection against the small arms projectiles is one of the most important needs of the nowadays military industry. Ballistic resistance of steel plates is complex. There are a lot of factors which have an impact on it. Ballistic resistance depends on target material behaviour, target thickness, angle of incidence, geometry and material of the projectile [7].Combat helmets have been used for many centuries to protect the most important part of the human body, which is the head. Contemporary warfare generates a lot of threats to which a soldier may be exposed. Based on the Body Surface Area (BSA) criterion, the head belongs to one of the most important parts of the human body and should be extremely protected. For this purpose, helmets are used which are characterized by resistance to ballistic strokes caused, among others, by: mines, fragments, or IED

© The Author(s), under exclusive license to Springer Nature Switzerland AG 2021
A. Mężyk et al. (Eds.): SMWM 2020, AISC 1336, pp. 258–268, 2021.
https://doi.org/10.1007/978-3-030-68455-6_23

(Improvised Explosive Device), as well as other incidents resulting directly from the soldier's actions in various terrain conditions [1, 2].

The design of a modern combat helmet is associated with several tests. The risk of head injury should be minimized. Modern ballistic helmets provide different levels of protection. The helmet is designed to protect the user's head against direct impact with fragments and some small arms. The helmet is also designed to protect the user's head against impact from objects falling from above, by resisting and deflecting blows to the head. Due to the low weight and high strength composite materials are more and more used to produce soldiers equipment. Combat helmets made from composites were a subject of much research. Paper [10] presents numerical and experimental ballistic tests of the composite combat helmet hit by the projectiles with various geometry. Paper [12] shows that, the impact resistance of the multilayered composite materials depends from the used reinforcing fabrics. Multilayered composite materials absorb higher amount of the impact energy when reinforcing fibers are made from more elastic material [9]. However, applied matrix material has also an impact to the ballistic resistance [4, 6]. In case of the composite materials manufacturing method has also an impact to the obtained mechanical properties [3].

The effectiveness of the protection of the soldier's head is determined by two basic criteria concerning, among others, helmets, they are:

- fragmentation resistance - i.e. the helmet's resistance to puncture with fragments,
- bulletproof - that is, the helmet's resistance to bullet penetration.

Table 1. Helmet fragmentation classes [8].

Fragmentation class	Ballistic protection limit V_{50} [m/s]	Deflection [mm]
O1	$400 \leq V50 \leq 500$	$Ug \leq 20$
O2	$500 < V50 \leq 600$	$Ug \leq 20$
O3	$V50 > 600$	$Ug \leq 20$

V_{50} ballistic protection limit is a projectile speed at which the probability of penetration of the test material is 50%. V_{50} is determined as the average of an equal number of the highest projectile velocities which cause only partial penetration and the lowest measured speeds of the projectile which results in complete penetration [8].

Table 2. Helmet bulletproof classes [8].

Bulletproofing class	Type of ammunition	Type of bullet	Bullet weight [g]	Bullet speed [m/s]	Number of bullet hits required
K1	357 Magnum	JSP	10.2	(381 ± 15) m/s	4
	9 × 19 mm	FMJ	8.0	(330 ± 15) m/s	4
K2	357 Magnum	JSP	10.2	(425 ± 15) m/s	4
	9 × 19 mm	FMJ	8.0	(360 ± 15) m/s	4

JSP – Jacketed Soft Point (semi-shell projectile with lead core)
FMJ – Full Metal Jacket (full-shell projectile with a jacket made of copper alloy)

The article compares the damage caused by the impact of the shrapnel and 9 × 19 mm projectile on a steel plate of different thickness and composite material as well. The conditions for which the plates meet the requirements of fragmentation and bulletproof were determined in accordance with PN-V-87001: 12011. The ballistic resistance classes of helmets in terms of fracture resistance are shown in Table 1, whereas the range of bullet resistance in Table 2 [8].

2   Fragmentation Resistant Analysis

2.1   Steel plate's Analysis

The numerical research was carried out using the finite element method and commercially available LS-Dyna software. Conducted research concerning the fragmentation was performed on the steel plates with various thicknesses and on the multilayered composite.

The article compares the damage caused by the impact of the shrapnel and then specifies the conditions for which steel plates meet the requirements of shrapnel and bullet resistance in accordance with PN-V-87001: 12011.

Model of the steel fragment was prepared in accordance to PN-V-87001: 2011 standard. Steel plate with dimension of 100 × 100 × 3 mm was modelled as a target. Fragment was placed in front of the target as it was shown in Fig. 1b. Such prepared model was discretised (created square elements were characterized by edge with length of 1 mm). As a boundary condition all degrees of freedom were fixed for the nodes located at the edges of the steel plate. Initial speed of 620 m/s was set for the fragment (in accordance with the third class of resistance [5]). The automatic surface to surface contact between the plate and the fragment was subsequently added.

Fragmentation resistance of 3, 4, 5 mm thick steel plates was compared. Location of the shrapnel at the beginning of simulation is presented in Fig. 1.

The material model that was used for steel plates considers the elasticity - plastic behaviour of the material. Both in the case of shrapnel and bullet resistance, standardized striking elements were used. In both cases, Johnson Cook's material model was used,

Fig. 1. Location of the projectile at the beginning of simulation.

which allows observation of the mushrooming process during impact. Johnson Cook material model parameters used in conducted research are shown in Table 3.

Table 3. Material parameters of steel plate and projectile [5, 7].

Variable	Unit	Steel plate/projectile
ρ	kg/m^3	7850
G	GPa	80
A	MPa	673
B	MPa	190
n	–	0.15
c	–	0.017
m	–	1.08
T_m	K	1800
T_R	K	298
ε_0	s^{-1}	0.0001
C_p	J/kgK	486
D_1	–	−0.19
D_2	–	0.732
D_3	–	0.663
D_4	–	0.029
D_5	–	0.716

In Figs. 2, 3 and 4 deformation of the 3–5 mm thick steel plates at different instances of the simulation are shown.

Results of numerical simulation for steel plates are presented in Table 4. Deformation of the 3, 4 and 5 mm thick steel plate at different instances of the simulation has shown

t = 0.006 ms   t = 0.0175 ms   t = 0.1 ms

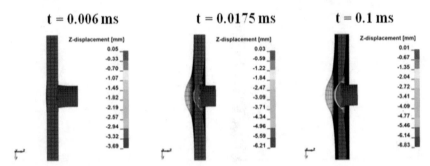

Fig. 2. Displacement of the 3 mm thick steel plate at different instances of the simulation.

t = 0.006 ms   t = 0.0175 ms   t = 0.1 ms

Fig. 3. Displacement of the 4 mm thick steel plate at different instances of the simulation.

t = 0.006 ms   t = 0.0175 ms   t = 0.1 ms

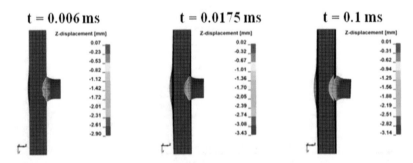

Fig. 4. Displacement of the 5 mm thick steel plate at different instances of the simulation.

that each of the steel plates passed the fragmentation resistance test, not exceeding 20 mm of displacement.

2.2 Epoxy/aramid Composite

Energy absorbing panels are more and more often made from composite materials. Those materials are characterized by low weight and high strength. Because of increasing popularity of the composite energy absorbing panels, authors decided to perform a numerical simulation of the shrapnel impact into 5 mm thick composite.

Table 4. Deflection of the steel plates after the impact.

Steel plate thickness [mm]	Fragmentation resistance
	Deflection [mm]
3	4.1
4	3.0
5	2.3

Modelling of the multilayered composite materials is much more complicated than in case of the isotropic material such as steel. There are three different scales in which composite material model could be prepared. Selection of appropriate modelling scale has a large impact on the obtained results. It is because, the damage of the composite material is a complex process. Depending on the used modelling scale, the numerical model would reflect energy-absorbing phenomenon, which occurs inside the composite material during the impact. The simplest to create is the model which treats the composite as a whole part – without dividing it into the reinforcing layers. In case of such modelling scale homogenization of the mechanical properties is conducted for the whole composite. Modelling in this scale makes it impossible to directly observe the delamination process. This is important, because delamination process absorbs a part of the impact energy. Second possibility of the composite material modelling considers division of the composite on the reinforcing layers. In this modelling scale each reinforcing layer is modelled as a different part. Connections between the reinforcing layers could be modelled by using a cohesive element or by the bilateral constraints (often with strength criterion). Observation of the delamination process and damage of the reinforcing material in each layer of the composite is possible in that modelling scale. The most advanced modelling scale of the multilayered composites considers the geometry of each reinforcing fiber. This approach reflects the phenomenon which occurs in the composite material. However, because of the high number of parameters which are necessary to define and high computational cost, the composite models prepared in this scale are rarely used (especially in complex models).

In the conducted research second of the mentioned modeling scale, which include division of the composite into the reinforcing layers, was used. Connections between reinforcing layers were modelled using the bilateral constraints with strength criterion (based on the normal and shear strength of the matrix).

The research was conducted for the composite material with 10 reinforcing layers. Fabric made from aramid fibers was used as a reinforcing material. Material model, which was selected for the reinforcing layers, is dedicated to the orthotropic materials. Mechanical properties of the epoxy/aramid composite which are used in the conducted research are presented in Table 5. In each reinforcing layer, X and Y axes were chosen as a direction of the reinforcing fibers.

Conducted research concerning fragmentation resistance showed that, the damaged formed in the composite create a characteristic cone. Area of damage increases with each successive reinforcing layer. Places where the connection between adjacent reinforcing

Table 5. Mechanical properties of epoxy/aramid composite [11].

Material properties	Value
Young's modulus $E_1 = E_2$ [MPa]	30000
Shear modulus G_{12} [MPa]	5000
Tensile strength $X_T = Y_T$ [MPa]	480
Compressive strength $X_C = Y_C$ [MPa]	190
Shear strength S [MPa]	50
Tensile strain $\varepsilon_{X_t} = \varepsilon_{Y_t}$ [%]	1.6
Compressive strain $\varepsilon_{X_c} = \varepsilon_{Y_c}$ [%]	0.6
Shear strain ε_S [%]	1
Density [g/cm^3]	1.4
Poisson's Ratio	0.2

layers is broken (delamination) are also visible. The picture of the composite damage presented in Fig. 5 showed that, impact resistance of the multilayered composite is much smaller than in case of the steel plate with the same thickness.

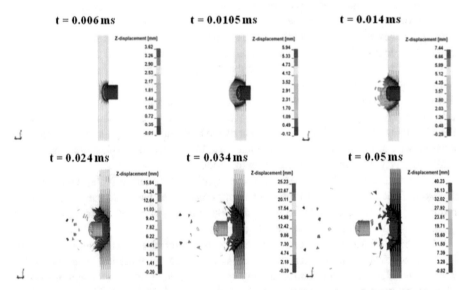

Fig. 5. Deformation of the epoxy/aramid composite at different time of the simulation.

Figure 6 shows a comparison of the kinetic energy of the shrapnel for the 5 mm steel plate and epoxy/aramid composite.

Low weight is the important feature of the multilayered composites which makes them so popular in case of energy absorbing panel application. Therefore, in the case of

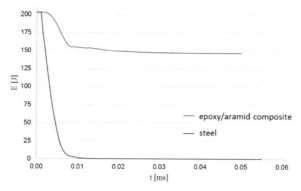

Fig. 6. Comparison of the kinetic energy of the projectile for the 5 mm steel plate and epoxy/aramid composite.

nowadays manufactured personal protective equipment, multilayered composite materials are often combined with a thin steel plate. This solution allows to achieve both, the low weight and the desired level of protection.

3 Steel plate's Analysis of Bullet-Proofing

The simulation was carried out at a projectile speed of 375 m/s, according to the second class of bullet-proofing [13]. Additionally, bullet-proofing of the 5 mm thick plate was checked when the projectile hit at higher speeds, until the plate was completely pierced. Location of the bullet at the beginning of simulation is presented in Fig. 7.

Fig. 7. Location of the bullet at the beginning of simulation.

Colored maps of the deformation after the impact of the projectile into steel plate with thickness of 3 mm, 4 mm and 5 mm is presented in Fig. 8.

Numerical simulation results of steel plates perforation with various thicknesses is presented in Table 6. On a base of results presented in Table 6 only a 5 mm thick plate passed the bullet resistance test.

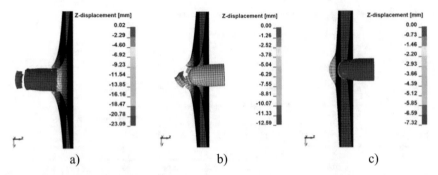

a)   b)   c)

Fig. 8. Colored maps of the displacement after the impact of the projectile into steel plate with thickness of a) 3 mm, b) 4 mm, c) 5 mm.

Table 6. Results of numerical simulation of bullet resistance.

Steel plate thickness [mm]	Bulletproof Perforation
3	YES
4	YES
5	NO

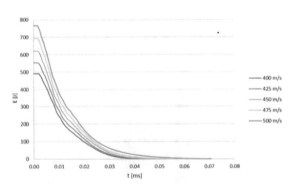

Fig. 9. Comparison of the kinetic energy of the projectile depending on initial speed of projectile for steel plate.

Figure 9 shows the kinetic energy of the projectile during the simulation at different initial speed while Table 7 compares the steel plate deflection depending on the projectile speed. Comparison of 5 mm steel plate deflection for the different bullet velocity is presented in Table 7. Ballistic limit for a 5 mm steel plate is about 500 m/s.

Table 7. Comparison of 5 mm steel plate deflection for the different bullet initial speed.

Bullet velocity [m/s]	Deflection [mm]
400	5.01
425	5.04
450	5.22
475	5.35
500	Perforation

4 Conclusion

Numerical tests showed that each of the tested steel plates underwent a fragment resistance test, not exceeding 20 mm deflection. Only 5 mm thick plate passed the bullet-proofing test, because it was not completely perforated. The ballistic limit for the testing steel plate is about 500 m/s.

Composite material with that same thickness as a steel plate is characterized by much smaller energy absorption during the impact. Solution in which the weight of the energy absorbing shield is minimized, and its desired level of protection is achieved, should contain a combination of multilayered composite with thin steel plate.

References

1. Jamroziak K.: Analiza hełmu balistycznego w ujęciu minimalizacji tępego urazu głowy analiza hełmu balistycznego w ujęciu minimalizacji tępego urazu głowy XIV Konferencja Naukowa, Majówka młodych Biomechaników" Im. Prof. Dagmary Tejszerskiej
2. Jamroziak K.: Próba oceny urazu głowy w ochronie balistycznej miękkiej. Modelowanie 42 s. **11**, 179–190 (2011)
3. Kyzioł, L., Panasiuk, K., Barcikowski, M., Hajdukiewicz, G.: The influence of manufacturing technology on the properties of layered composites with polyester–glass recyclate additive. Prog. Rubber Plast. Recycl. Technol. **36**(1), 18–30 (2020)
4. Mayer, P., Pyka, D., Jamroziak, K., Pach, J., Bocian, M.: Experimental and numerical studies on ballistic laminates on the polyethylene and polypropylene matrix. J. Mech. **35**(2), 187–197 (2019)
5. Özel, T., Karpat, Y.: Identification of constitutive material model parameters for high-strain rate metal cutting conditions using evolutionary computational algorithms. Mater. Manufact. Process. **22**(5), 659–667 (2007)
6. Pach, J., Mayer, P., Jamroziak, K., Polak, S., Pyka, D.: Experimental analysis of puncture resistance of aramid laminates on styrene-butadiene-styrene and epoxy resin matrix for ballistic applications. Arch. Civil Mech. Eng. **19**(4), 1327–1337 (2019)
7. Palta, E., Hongbing, F., Weggel, D.C.: Finite element analysis of the advanced combat helmet under various ballistic impacts. Int. J. Impact Eng **112**, 125–143 (2018)
8. PN-V-87001:2011. Osłony balistyczne lekkie – Hełmy ochronne odłamko- i kuloodporne – Wymagania i badania
9. Reddy, T.S., Reddy, P.R.S., Madhu, V.: Response of E-glass/epoxy and Dyneema® composite laminates subjected to low and high velocity impact. Procedia Eng. **173**, 278–285 (2017)

10. Rubio, I., Rodríguez-Millán, M., Marco, M., Olmedo, A., Loya, J.A.: Ballistic performance of aramid composite combat helmet for protection against small projectiles. Compos. Struct. **226**, 111153 (2019)
11. Sławski, S., Szymiczek, M., Kaczmarczyk, J., Domin, J., Duda, S.: Experimental and numerical investigation of striker shape influence on the destruction image in multilayered composite after low velocity impact. Appl. Sci. **10**(1), 288 (2020)
12. Sławski, S., Szymiczek, M., Kaczmarczyk, J., Domin, J., Świtoński, E.: Low velocity impact response and tensile strength of epoxy composites with different reinforcing materials. Materials **13**(14), 3059 (2020)
13. STANAG 4090, Small arms ammunition (9 mm parabellum)

Structural Analysis of 6R Robotic Arm. Comparison of Different Complexity Models

Michał Soida[1]([⊠]) [iD], Jakub Żak[1] [iD], and Sławomir Bydoń[2] [iD]

[1] AGH University of Science and Technology, Al. Mickiewicza 30, 30-059 Kraków, Poland
michal@soida.pl, kontakt@jakubzak.pl
[2] Multiprojekt Automatyka sp. z o.o., ul. Fabryczna 20a, 31-553 Kraków, Poland
sbydon@multiprojekt.pl

Abstract. Industrial manipulators are replacing manual workers in an increasing number of professions. Robots can nowadays be found in nearly all sectors of industry. Therefore, in an industry dominated by big companies (KUKA, ABB, FANUC), more smaller companies are trying to design their own robotic manipulators. Unfortunately, this is not an easy task – in addition to a well-equipped machine park, there is a need for scientific and research facilities, which small and medium companies cannot afford. The first step of structure analysis is to choose a proper kinematic and dynamic structure. In general, it is a very complex issue, and therefore there is doubt as to how simplified model can be adopted so that the obtained results are reliable. According to the authors' knowledge, there is currently no research available that would clearly indicate the validity of introducing such simplifications to the model. In this work, the most popular robot model was considered – a 6-axis manipulator with an open kinematic chain. The selected structure was based on a market analysis for small and medium-sized manipulators. In each subsequent step, additional details were added to the manipulator structure and the results were compared to each other. The conclusion of this work shows that increasing the complexity of the model is justified, but it only makes sense to a certain point. Further refinement significantly increases the time it takes to carry out the analysis, only slightly affecting the accuracy of the results.

Keywords: Manipulating robot · Kinematic analysis · Dynamic analysis

1 Introduction

The last decade has shown a significant increase in the robotics sector. Nowadays robots can be found in almost all branches of economy, from high-tech factories to modern life. This phenomenon is explained by the rapidly developing trend aimed at minimizing human labor, especially in places where it is dangerous for human health and life [1]. Advanced robotics allow for partial independence from the human factor, as human behavior cannot be accurately predicted. Recent events in the world, such as the pandemic caused by SARS-CoV-2 (1st quarter of 2020), stress even more how important is the development of robotics.

A. Mężyk et al. (Eds.): SMWM 2020, AISC 1336, pp. 269–278, 2021.
https://doi.org/10.1007/978-3-030-68455-6_24

International Federation of Robotics (IRF) is a professional non-profit organization that aims to promote research and development in the field of robotics. For years, IRF has been collecting statistical data on the demand and sales of robots in the world. Since 2011, the demand for industrial robots has been constantly increasing (see Fig. 1), and thus the robotization ratio has increased in many countries.

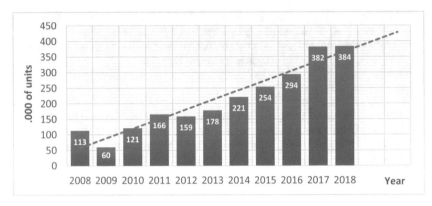

Fig. 1. Estimated worldwide annual supply of industrial robots [2].

Currently, the industrial robot market is heavily dominated by several large companies. In accordance to data from 2019 [3], Fanuc, ABB, KUKA and Yaskawa have more than half of the market share (57%). However, the constantly increasing demand is a big opportunity for small and medium companies to get to market with their own designs. Robotics, due to its multidisciplinary nature [1, 4], is big challenge for the new companies in the industry. The development of such devices requires both industrial practice and theoretical analysis. The case study shows that lack of theoretical workshop is the only reason that prevents companies from getting involved in the robotic industry.

In this paper, the authors take up the preparation of a calculation procedure that allows the calculation of the required manipulator drive moments. Various approaches to the topic have been analyzed and a design procedure has been developed.

2 Problem

The company in which the authors work decided to join the group of companies involved in the production of industrial robots. In 2018, a decision was made to design three 6-axis manipulators. One educational, with the features of a typical industrial robotic arm, intended for the market of universities [5] and the other two strictly for the industrial sector.

An analysis of the selection of drive system parameters is required for each of the robots separately. Unfortunately, no general guidelines are available to determine the choice of method or the accuracy of the mathematical model. Thus, the authors decided to find the answer to the following research problem:

What impact does the quality of the adopted mathematical model have on drive torque calculation?

3 Robotic Arm Design

The industrial manipulator is a serial mechanical system design, in terms of shape similar to the human upper limb [6]. Referring to a standard [7], it is an automatically controlled, programmable in three or more axes device, that can be both mobile or stationary, dedicated to industrial automation.

Several basic kinematic structures are widespread in industry. They differ in the types of joint used (P-prismatic joint, R-swivel joint) and their configuration. Each structure is characterized by its typical application and work volume [6, 8, 9] (see Table 1).

Table 1. Structure type and typical use of the manipulator.

Type	cartesian	cylindrical	SCARA	anthropo-morphic	spherical
Structure	P-P-P	R-P-P	R-R-P	R-R-R	R-R-P
Application	transportation	palletizing	assembly pick&place sorting	assembly pick&place painting polishing	assembly painting
Work space					

The authors decided to choose an anthropomorphic structure. The main reason for choosing was compliance with the assumptions of the company project.

Market analysis (see Table 2) shows that small manipulators (range up to 1 m and load capacity up to 6 kg) are based on the same general kinematic scheme (see Fig. 2).

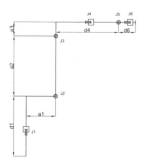

Fig. 2. The general kinematic structure of 6R manipulator.

Additionally, analyzing the figures (see Table 2), one can come to the following conclusions:

1. Most of the analyzed manipulators have a non-zero offset (a1) between motion axes J2 and J3. This solution is advantageous. The range and workspace is greatly increased without any impact on the J2–J6 drive units.
2. Most of the analyzed manipulators have a non-zero offset (a3) between motion axes J3 and J4. This solution facilitates wiring to the wrist module.
3. The existence of the mentioned offsets allows to eliminate some singular positions.

Table 2. 6R robotic arms overview. Variable names with accordance to Fig. 2. Data from respective robots' datasheets.

Brand/model	d1	a1	a2	a3	d4	d6
Yaskawa GP7	330	40	385	???	340	80
KUKA KR6 AGILUS SIXX R900	400	25	455	35	420	80
COMAU RACER	430	150	447	130	392	100
MITSUBISHI MELFA RV-7FL	400	0	435	50	470	85
ABB IRB 120	290	0	270	70	302	72
ABB IRB 1200	400	0	448	42	451	82
STAUBLI TX60	375	0	290	0	310	70
ABB IRB 140	362	70	360	0	380	65
FANUC LR MATE 200id	330	75	300	75	320	80
OMRON ADEPT VIPER S850	355	75	365	90	405	80
KAWASAKI RS005L	295	105	380	80	410	78

4   Methodology

The main purpose of the analysis is to determine drive unit parameters for each of the robotic arm axes. Drive torque in general is expressed as a sum of static and dynamic components.

$$T = T_s + T_d \tag{1}$$

$$T_s = m \cdot x_{com} \cdot g \tag{2}$$

$$T_d = I \cdot \varepsilon \tag{3}$$

To prepare the manipulator structure for calculations, homogeneous transformation was used. Each joint in the considered robot is a class 5 kinematic pair, which means it has only 1 degree of freedom. Given this, by far the most favorable form of manipulator movement description is the use of Denavit-Hartenberg notation [10]. DH notation assumes the use of a relationship (rotation and translation) to describe the motion with respect to only two axes: X and Z (4). Each formula (4) must contain only one variable parameter.

$$A_i = Rot_{z,\Theta i} \cdot Trans_{z,di} \cdot Trans_{x,ai} \cdot Rot_{x,\alpha i} \qquad (4)$$

$$T_{0,n} = A_0 \cdot A_1 \cdot \ldots \cdot A_n \qquad (5)$$

According to these guidelines, the coordinate system is moved to the next key position inside the manipulator structure (see Fig. 3).

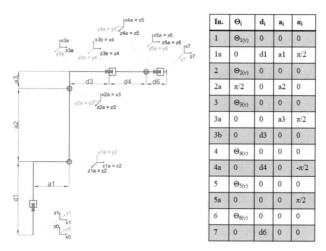

In.	Θ_i	d_i	a_i	α_i
1	$\Theta_{1(v)}$	0	0	0
1a	0	d1	a1	$\pi/2$
2	$\Theta_{2(v)}$	0	0	0
2a	$\pi/2$	0	a2	0
3	$\Theta_{3(v)}$	0	0	0
3a	0	0	a3	$\pi/2$
3b	0	d3	0	0
4	$\Theta_{4(v)}$	0	0	0
4a	0	d4	0	$-\pi/2$
5	$\Theta_{5(v)}$	0	0	0
5a	0	0	0	$\pi/2$
6	$\Theta_{6(v)}$	0	0	0
7	0	d6	0	0

Fig. 3. Kinematic structure and DH table for the considered manipulator. (v) means variable.

Referring to the formula (3), to find the dynamic torque a proper acceleration value must be assumed. Incorrect assumption will lead to inaccurate results. It is possible to choose the acceleration value based on the selected components of the drive system. However, at this stage of the design, the exact components have not yet been selected. Moreover, manufacturers of the manipulators do not provide such data in their catalogues.

However, the standard of robot safety [11] requires the manufacturers to provide values of stopping time and stopping distance for three significant axes of motion when invoking STOP commands specified by [12]. Using the data for STOP1 (a controlled stop with power available to the machine actuators to achieve the stop, and then the removal of power when stop is achieved), it is possible to estimate the value of delays for commercially available devices (see Table 3).

According to the received data, the adopted acceleration value should not be higher than 3.9 rad/s^2.

Table 3. Stopping time and distance of Omron Viper s650 [13] and Kuka KR Agilus R700 sixx [14].

Axis	Stopping distance [deg]		Stopping time[s]		Deceleration [rad/s^2]	
Robot	Omron	KUKA	Omron	KUKA	Omron	KUKA
Axis 1	55	70	0,3	0,33	−3,2	−3,7
Axis 2	65	60	0,41	0,33	−2,8	−3,2
Axis 3	72	58	0,38	0,26	−3,3	−3,9

5 Experiment

To answer the questions posed by the research hypothesis, two refinement directions were considered – gradually increasing realism of chosen manipulator structure (first point masses, then bars with negligible cross-section, and finally cylinders), and extracting some elements from the homogenous parts – i.e. motors and gearboxes, since they constitute a significant part of the overall mass (Fig. 4).

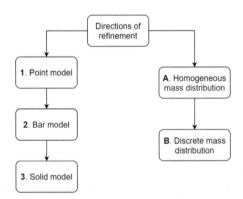

Fig. 4. Directions of refinement.

This strategy allows to consider effects of these refinements both in relation to their mathematical complexity, and how they affect accuracy gains of each other.

The first direction of refinement (complexity of mathematical model) was categorized into three steps. First, the simplest point mass model was created (see Fig. 3). This allows for minimal mathematical complexity and requires only basic model assumptions – general dimensions and mass of the manipulator parts – for the Steiner's theorem.

The second step adds the first dimension to manipulator elements – i.e. the length of robot links – making it a bar model. This still keeps the required assumptions at a minimum, only adding some complexity to calculations – since the length of links is already required to calculate the center of mass. The third step adds a second dimension, and a new assumption – thickness of the links. In this analysis, all links were assumed to have a circular cross-section with a diameter of 100 mm.

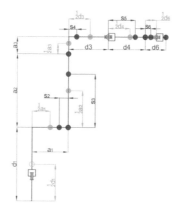

Fig. 5. Kinematic structure with discrete mass distribution.

The second direction of refinement – extracting gearboxes and motors – might look much easier to achieve, since it only adds additional point masses. (see Fig. 5) However, it requires a lot more assumptions about the manipulator structure – so that precise positions of each motor and gearbox can be obtained. This makes it impossible to perform such analysis before having at least a first draft design of a concrete manipulator structure design. In our analysis, all gearboxes were assumed to be placed at their respective joints, and motors were offset.

To perform the analysis, open-source software JupyterLab [15] was used, which allows to create interactive calculation scripts in the Python programming language. For symbolic computation, the SymPy [16] library was used – thanks to this, one set of transformations can be used for all considered cases. A visualization of calculated geometric points (for graphical confirmation of used transformations) was created with the help of MatPlotLib plotting library.

The Analysis was Performed According to This Algorithm

1. Definition of all symbolic variables
2. Definition of all input values
3. Definition of kinematic transformation model
4. Calculation of the static torque (cases A and B)
5. Calculation of the moment of inertia (cases 1, 2, 3)
6. Calculation of the dynamic torque (both A/B and 1/2/3 – 6 cases total)
7. Substitution of symbolic variables with input values, displaying results

For each considered joint, the manipulator was set in a worst-case pose – for joints 1, 2, 3 and 5 this means maximally extended and parallel to the ground. To create such worst-case pose for joint 4, which rotates on a different plane, joint 5 was rotated by 90°.

A big benefit of the analysis described above is the possibility of simple adjustment to other robots. For anthropomorphic six-axis manipulators, only input values have to be adjusted (p. 2). For other structures, kinematic transformations have to be modified (p. 3).

6   Results

6.1   Model Type Comparison (Case 1A, 2A, 3A)

Table 4. Dynamic torque in model type comparison. (Torque in Nm).

	1A	Δ_{12}	2A	Δ_{23}	3A	Δ_{13}
J1	22,074	1,50%	22,405	0,25%	22,461	1,76%
J2	13,498	2,45%	13,829	0,41%	13,885	2,87%
J3	2,997	1,73%	3,048	0,83%	3,074	2,57%
J4	0,011	**33,64%**	0,014	**44,06%**	0,021	92,52%
J5	0,011	**33,64%**	0,014	**44,06%**	0,021	92,52%

Table 5. Drive torque in model type comparison. (Torque in Nm).

	1A	Δ_{12}	2A	Δ_{23}	3A	Δ_{13}
J1	22,074	1,50%	22,405	0,25%	22,461	1,76%
J2	87,467	0,38%	87,798	0,06%	87,854	0,44%
J3	29,245	0,18%	29,297	0,09%	29,322	0,26%
J4	0,769	0,47%	0,772	0,82%	0,778	1,29%
J5	0,769	0,47%	0,772	0,82%	0,778	1,29%

There is a significant dynamic torque change in J4 and J5 (max. ~44%). However, due to static load being much higher, this does not create a noticeable difference in overall drive torque (Tables 4 and 5).

6.2   Mass Type Comparison (Case 1A and 1B)

In this case, there are noticeable changes in all joints, especially in J4 and J5 – both in static and dynamic torque (Table 6).

6.3   Model and Mass Comparison (Case 1B, 2B, 3B)

In this case, effects of structure representation change are negligible (~5%) (Tables 7 and 8).

Table 6. Static torque, dynamic torque and drive torque in mass type comparison. (Torque in Nm).

	1A dynamic	$\Delta_{12\ dynamic}$	1B dynamic	1A static	$\Delta_{12\ static}$	1B static	1A drive	$\Delta_{12\ drive}$	1B drive
J1	22,07	6,1%	23,4	0,0	-	0,0	22,07	6,0%	23,40
J2	13,49	8,9%	14,7	73,96	6,6%	78,9	87,46	7,0%	93,62
J3	2,99	12%	3,36	26,24	2,7%	26,97	29,24	3,7%	30,33
J4	0,01	**135%**	0,025	0,758	**46%**	1,10	0,769	**47%**	1,13
J5	0,01	**135%**	0,025	0,758	**46%**	1,10	0,769	**47%**	1,13

Table 7. Dynamic torque in model mass comparison. (Torque in Nm).

	1B	Δ_{12}	2B	Δ_{23}	3B	Δ_{13}
J1	23,408	0,82%	23,599	0,12%	23,6283	0,94%
J2	14,7109	1,30%	14,9019	0,20%	14,9312	1,50%
J3	3,3624	0,73%	3,3869	0,32%	3,3977	1,05%
J4	0,0252	3,57%	0,0261	5,75%	0,0276	9,52%
J5	0,0252	3,57%	0,0261	5,75%	0,0276	9,52%

Table 8. Drive torque in model mass comparison. (Torque in Nm).

	1B	Δ_{12}	2B	Δ_{23}	3B	Δ_{13}
J1	23,408	0,82%	23,599	0,12%	23,628	0,94%
J2	93,627	0,20%	93,818	0,03%	93,847	0,24%
J3	30,335	0,08%	30,359	0,04%	30,370	0,12%
J4	1,130	0,08%	1,131	0,13%	1,133	0,21%
J5	1,130	0,08%	1,131	0,13%	1,133	0,21%

7 Conclusion

As it can be seen in the previous chapter, changing the mass distribution has the biggest impact on the calculated values. Therefore, extracting motors and gearboxes from the homogenous manipulator structure is definitely recommended. This should be done before considering more advanced structure models which make the mathematical model much more complicated, giving only small result accuracy benefits.

The second important conclusion is that the complexity of the structural model is only relevant for inertia calculations – its influence on the static load is negligible. In most cases, after separating motors and gearboxes, refining the material point model has no benefits – however, it can be useful in joints 4 and 5, where the distances are smaller, and the moment of inertia plays a much bigger role in the received results.

8   Future Plans

Using the above analysis, the authors plan to develop software that enables automatic computation of necessary parameters of a robot drive. The analysis will be extended to allow calculations for any robot pose (user-supplied values for all manipulator joints). In the next step, the algorithm will be extended for other manipulator types – e.g. SCARA – allowing the tool to be used for most typical industrial robot designs.

At the same time, the results received from this analysis will be compared to a fully developed multibody simulation, on a detailed CAD model, using dedicated software – MSC Adams. In the future, the results will be compared to a real, physical manipulator model.

Acknowledgements. This research was financially supported by the Project no. RPMP.01.02.01-12- 0494/16-00.

References

1. Honczarenko, J.: Roboty Przemysłowe. Budowa i zastosowanie. Wydawnictwo Naukowo-Techniczne, Warszawa (2004)
2. International federation of robotics – statistics. https://ifr.org. Accessed 08 May 2020
3. Industrial robots market shares. https://www.statista.com/statistics/317178/leading-industrial-robot-companies-globally-by-revenue. Accessed 08 May 2020
4. Zdanowicz, R.: Podstawy Robotyki. Wydawnictwo Politechniki Śląskiej, Gliwice (2011)
5. Bydoń, S., Żak, J., Soida, M., Macioł, J., Lisak, R.: Prototype of a six-axis educational manipulator. Projektowanie i dynamika urządzeń Mechatronicznych: zagadnienia wybrane, pp. 61–68 (2019)
6. Morecki, A., Knapczyk, J., Galicki, M., Marszalec, J., Wiliński, A., Zielińska, T.: Podstawy Robotyki. Teoria i elementy manipulatorów i robotów. Wydawnictwo Naukowo-Techniczne, Warszawa (1999)
7. ISO 8373:2012: Robots and robotic devices - Vocabulary
8. Division and use of manipulators. https://www.asimo.pl/teoria/roboty-przemyslowe.php. Accessed 08 May 2020
9. Division and use of manipulators. https://www.robotyka.com/teoria_spis.php. Accessed 08 May 2020
10. Buratowski, T.: Podstawy Robotyki. Uczelniane Wydawnictwo Naukowe, Kraków (2006)
11. ISO 10218-2:2011: Robots and robotic devices—safety requirements for industrial robots—part 1: robots
12. IEC 60204-1:2016: Safety of machinery - electrical equipment of machines - part 1: general requirements
13. OMRON Viper s650 datasheet. https://industrial.omron.pl/pl/products/viper. Accessed 08 May 2020
14. KR Agilus R700 sixx datasheet. https://www.kuka.com/en-us/products/robotics-systems/industrial-robots/kr-agilus. Accessed 08 May 2020
15. Project Jupyter. https://jupyter.org. Accessed 08 May 2020
16. SymPy documentation. https://www.sympy.org/en/index.html. Accessed 08 May 2020

Finite Element Analysis of Stress Distribution in the Node Region of Isogrid Thin-Walled Panels

Łukasz Święch$^{(\boxtimes)}$ (iD)

Faculty of Mechanical Engineering and Aeronautics, Rzeszów University
of Technology, al. Powst. Warszawy 8, 35-959 Rzeszow, Poland
lukasz.swiech@prz.edu.pl

Abstract. The paper presents numerical investigations of thin-walled integrally stiffened plates subjected to in-plane loading. Stiffening was in the form triangle grid called isogrid. Analytical solutions, on account of geometrical complexity of such structures, are mainly based on smeared stiffener and are limited only to preliminary design phase. To obtain more accurate results the application of FEM is required. The aim of the study was to determine the discretization strategy, type of used finite elements and its number on the accuracy of numerical results. Numerical representation of the considered structure was developed with the use of ABAQUS commercial FEM code. Each of the models was analyzed with taking into account both geometrical and physical nonlinearities. In particular, the focus was aimed at the analysis of the stress distribution in the area of the node of the grid. The conducted research allowed to compare numerical results among themselves and in relation to experimental investigations conducted with the use of the DIC scanner. The obtained results made it possible to formulate recommendations regarding possible simplifications of numerical modelling of the considered isogrid structures.

Keywords: Post-critical state · Grid stiffened structure · Finite element analysis

1 Introduction

The development of production methods such as numerically controlled machine tools allows the creation of so-called integral constructions, in which both the skin and stiffeners of the thin-walled structure are made of one piece of material. Integral grid stiffening is one of the possibilities to reduce the structural weight while maintaining strength and stiffness of the construction. Such an approach makes it possible to reduce the weight of the structure by eliminating elements connecting individual parts. To date, integral structures are used primarily in the aerospace industry.

Stiffened structures called isogrid, consist of ribs and skin arranged in the uniform pattern of triangles and may be treated as layered material, with appropriate elastic constants for each element [1]. Analytical solutions, on account of geometrical complexity, are mainly based on smeared stiffener methods which are constantly under development

A. Mężyk et al. (Eds.): SMWM 2020, AISC 1336, pp. 279–288, 2021.
https://doi.org/10.1007/978-3-030-68455-6_25

[2–4]. The method enables determination of the stiffness properties of the analyzed elements and solution procedures are based directly on the theory of elasticity [5, 6] and can be found in numerous publications related to plates analyses [7–14]. This approach allows to identify critical and limit loads for isogrid structures, but it is limited to the preliminary design phase. To obtain more accurate results, especially in the range of post-critical, geometrically nonlinear deformations, application of FEM is required [15–17]. Complication in grid geometry also leads to the risk of occurrence high stress concentrations areas which can have a significant impact on fatigue life of the structure [18–20].

The paper presents numerical investigations of post-buckling behavior of thin-walled plates integrally stiffened by a triangular pattern of ribs subjected to pure shear loading conditions. The aim of the study was to determine the influence of the FEM discretization method on the accuracy of numerical results in comparison to experimentally obtained with the use of DIC system. Apart from the analysis of the convergence of the numerical solution concerning the global deformation of the structure, the influence of the size of the finite elements on the stress distribution in the node region of isogrid was also assessed, assuming that the accuracy of the numerical representation of this quantity may have a decisive influence on the correctness of fatigue life predictions.

2 Experimental Investigations

Experimental tests were carried out on the 306 × 306 mm square plate with a test field of 275 × 275 mm. The plate was machined from 2 mm thick duralumin 2024 T3 plate with the use of CNC milling machine and specially prepared vacuum table. Such a test structure was prepared as an integrally stiffened plate with periodic isosceles triangle grid of stringers of dimensions 45 mm by 90 mm (Fig. 1a). Stringers were 2 mm thick and 1.5 mm high. In every node in the grid the hole with diameter of 3 mm was drilled. The skin of structure, areas between stringers, had a thickness of 0.5 mm (Fig. 1b).

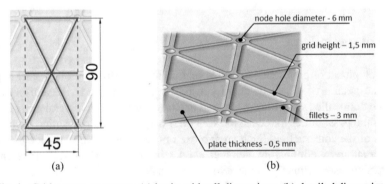

Fig. 1. Grid pattern geometry: (a) basic grid cell dimensions, (b) detailed dimensions.

The plate was mounted in the steel frame enabling converting vertical force into the pure shear loading condition (Fig. 2). Experimental tests were conducted with the use of strength testing machine ZWICK Z050 with force level control during the test. Deformations were measured by 3D DIC (Digital Image Correlation) system ARAMIS [21, 22]. Connection between testing devices allowed to capture spatial field of deformation with information of acting force in every step of loading. Figure 3 presents photography of plate deformation and deflection fields of the plate captured by DIC system at the level of maximum load. Experimental results used as the reference solution were widely described in [23].

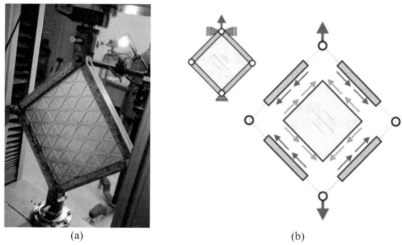

(a)   (b)

Fig. 2. Experimental setup: (a) plate mounted in the test area of the strength machine, (b) loading conditions.

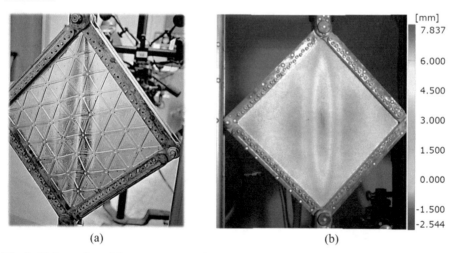

(a)   (b)

Fig. 3. Deformation of the structure (a) and deflection field captured by DIC system at the level of maximum load (b).

3   Numerical Analyses

In order to compare with the results of the experiment, numerical model of the considered structure was developed with the use of ABAQUS commercial FEM code.

The geometry of the plate was discretized with use of S4R and S3 shell elements with different thickness for areas of plate and stiffeners. Steel frame was modelled by beam B31 elements. Frame members were connected in the corners by means of pivot connection type. Plate and frame elements were connected with the use of the TIE constraint. To reproduce experimental boundary conditions, a sliding and fixed supports was applied at the upper and bottom node of the frame, respectively. Figure 4 presents the numerical FEM model and boundary conditions in case of the model with characteristic element length equal to 2.5 mm.

The material of the frame was perfect linear elastic and for the plate, bilinear elasto-plastic approximation of material model was used. Young modulus of aluminium alloy for elastic range was E = 70 GPa, Poisson ratio was $v = 0.33$, the yield point was Re = 345 MPa. For the plastic range of strains tangent modulus was $E_T = 0.9$ GPa and the ultimate stress point had a value of Ru = 488 MPa at the point of plastic strain equal to 0.164.

Numerical analyses were carried out with the use of Newton-Raphson procedures for nonlinear FEM solution to allow nonlinear geometric and plastic deformations [24–26].

A convergence study of the post-buckling behavior of the structure was performed to define the relation between mesh size and accuracy of results in comparison to experimental investigations. Figure 5 presents the element mesh grids used in the convergence study.

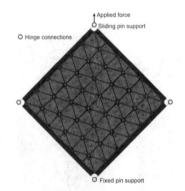

Fig. 4. FEM model and boundary conditions.

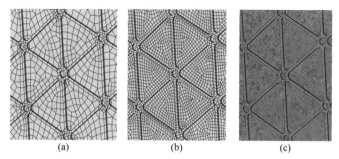

(a)   (b)   (c)

Fig. 5. FEM model and boundary conditions and characteristic element length a) 5 mm, b) 2,5 mm, c) 1 mm.

4 Comparison of Results

Results obtained by the numerical analyses were compared in two aspects. The first was a comparison with the results of experimental research to determine the correctness of the results obtained. The second comparative aspect was the study of the effect of the size of elements on the result in both the form of deformation and the state of stress in the examined structure.

The numerically predicted global, in-plane stiffness (Fig. 6) was in good agreement with the experiment to a load level of about 75 N/mm. Above that value of the acting load, the FEM model had a higher in-plane stiffness for every considered numerical model. It must be noted that at a load level of 36 N/mm, the numerical results revealed initial plastic deformation of the structure.

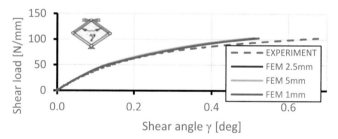

Fig. 6. Comparison of results of shear load versus shear angle obtained from FEM for the stiffened plate in the convergence study.

The second chosen so-called equilibrium path for validation of numerical modelling correctness in relation to the experimental results was relationship between the acting load and deflection of the point in the geometrical center of the plate (Fig. 7). It can be found that in every step of loading between the level of about 17 N/mm and 75 N/mm, FEM models were less stiff in out-of-plane direction related to deflection of the structure. Differences are can be explained by simplifications in the geometry and model material used in numerical simulations.

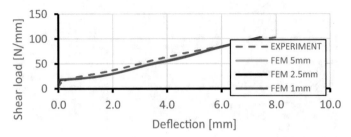

Fig. 7. Comparison of results of shear load versus deflection at the geometrical center of the plate obtained from FEM for the stiffened plate in the convergence study.

Comparison of results obtained for models with different element sizes leads to the conclusion that also in this aspect, characteristic element size has a low influence on stability prediction of the structure.

As mentioned, besides analysis of results in light of the experimental investigation, FEM models with different mesh size have been compared among themselves.

Figures 8, 9 and 10 present results of numerical calculations in the form of deflection fields and HMH stress distribution printed for the stage of the maximum acting load, for models with characteristic finite element size equal to 5 mm, 2.5 mm and 1 mm respectively.

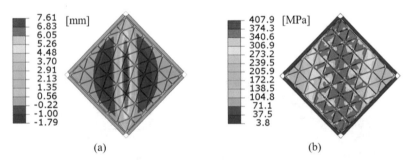

(a)   (b)

Fig. 8. Deflection field (a) and HMH stress distribution (b) for the model with 5 mm elements length.

In the case of the deflection fields, differences between maximum values obtained for models under consideration did not exceed 1% and can be neglected. Highest values of deflections were observed for the model with 1 mm elements, what certainly was related to the highest level of stress obtained for this model (Fig. 10b). Differences in stress levels between models were much higher than for deflections and dissimilarity between models with 1 mm and 5 mm elements was 13%. It is worth to noted that with decrease in the element size, stress concentrations rise in the area near upper and lower nodes of the mounting frame. Figures 11 presents an enlarged view of stress distribution and stress concentration indicated above for models with 5 mm and 1 mm element length.

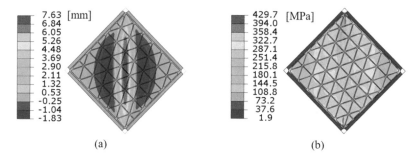

Fig. 9. Deflection field (a) and HMH stress distribution (b) for model with 2,5 mm elements length.

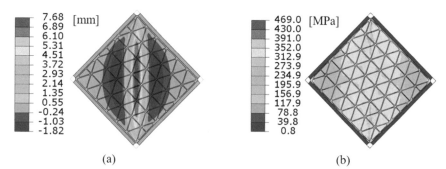

Fig. 10. a) Deflection field (a) and HMH stress distribution (b) for the model with 1 mm elements length.

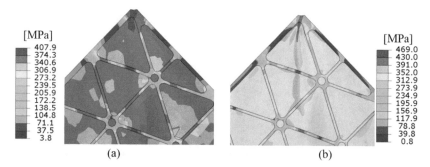

Fig. 11. Comparison of HMH stress states in the area of upper connection node of the model for maximal loading, for models with element characteristic length: a) 5 mm, b) 1 mm.

To better illustrate the relation between element size and stress distribution around the nodes of the grid in Figs. 12 and 13 present HMH stress states for the stage of deformation, before the loss of stability of the structure (shear about 17 N/mm) and for the maximum level of load respectively, for models with highest and lowest mesh density. Location of node presented in these figures was chosen in the center of the plate to present area affected by boundary conditions as little as possible.

Besides obvious differences related to the size of used elements and connected with that possibilities to simulate local effects, it should be noted that in the considered FEM models did not reproduce fillets between stringers and skin, which probably has an impact on stress levels in real construction.

Based on conducted researches and numerical analyses, some conclusions can be formulated. First of all, the size of the mesh has a small influence on the overall stiffness of the structure under consideration in the static load conditions. H-M-H stress distribution obtained for models with different characteristic element length (Fig. 6) varied significantly, which can have a decisive impact on the estimation of fatigue life.

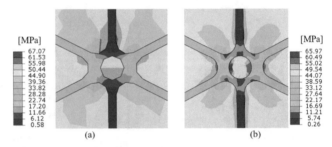

Fig. 12. Comparison of H-M-H stress states in the node of the stiffening grid for load about 17 N/mm, related to pre-buckling state of deformation. Results for models with element characteristic length equal to 5 mm (a) and 1 mm (b).

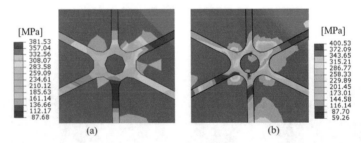

Fig. 13. Comparison of H-M-H stress states [MPa] in the node of the stiffening grid for maximal loading related to post-buckling state of deformation. Results for models with element characteristic length equal to 5 mm (a) and 1 mm (b).

5 Conclusions

Based on numerical analyses validated by conducted experimental research, some conclusions can be formulated. Both, experimental and numerical analyses lead to the convergence results in the case of the shape and values of the global deformation. Observed differences had potential reasons in simplified FEM model, especially in the absence of fillets between plate and stringers, simplified material model, initial geometrical imperfection.

The numerical analyzes carried out also prove that size of the mesh has a small influence on the overall stiffness of the structure under consideration in the static load conditions. Representative deflections and shear angles for models with characteristic element length varied from 1 mm to 5 mm did not indicate significant differences.

Contrary to the above, H-M-H stress distribution obtained for the same numerical models and loading conditions varied significantly. The analyzes carried out indicate that special attention should be placed on the area of stress concentrations, which were observed in the vicinity of the isogrid nodes. For analysed models differences in stress levels connected with used element size reaches 13% between models with 1 mm and 5 mm characteristic length. This phenomenon could have a decisive impact on the estimation of the fatigue life of the structure, but it should be noted that for this purpose, more precise numerical analyzes should be carried out, enabling the most accurate mapping of the geometrical details of the considered structures.

References

1. Isogrid Design Handbook. Report, CR-124075. NASA Marshall Space Flight Center, Huntsville (1973)
2. Luan, Y., Ohlrich, M., Jacobsen, F.: Improvements of the smearing technique for cross-stiffened thin rectangular plates. J. Sound Vibr. **330**, 4274–4286 (2011)
3. Wang, D., Abdalla, M.: Global and local buckling analysis of grid-stiffened composite panels. Compos. Struct. **119**, 767–776 (2015)
4. Xu, Y., Tong, Y., Liu, M., Suman, B.: A new effective smeared stiffener method for global buckling analysis of grid stiffened composite panels. Compos. Struct. **158**, 83–91 (2016)
5. Timoshenko, S., Gere, J.: Theory of Elastic Stability. Dover Publications, Mineola (2009)
6. Bauchau, O., Craig, J.: Structural Analysis: With Applications to Aerospace Structures. Springer, Dordrecht (2009)
7. Szilard, R.: Theory and Analysis of Plates: Classical and Numerical Methods. Prentice-Hall, Englewood Cliffs (1973)
8. Kączkowski, Z.: Plates, Static Calculations (in Polish). Arkady, Warsaw (1980)
9. Timoshenko, S., Woinowsky-Krieger, S.: Theory of Plates and Shells. McGraw-Hill, New York (1987)
10. Reddy, J.N.: Theory and Analysis of Elastic Plates and Shells. CRC Press, Boca Raton (2007)
11. Ugural, A.: Stresses in Beams, Plates, and Shells. CRC Press, Boca Raton (2009)
12. Bloom, F.: Handbook of Thin Plate Buckling and Postbuckling. Chapman & Hall/CRC, Boca Raton (2000)
13. Doyle, J.F.: Nonlinear Analysis of Thin-Walled Structures. Springer, Berlin (2001)
14. Vetyukov, J.: Nonlinear Mechanics of Thin-Walled Structures. Springer , Vienna (2014)

15. Hilburger, M., Lindell, M., Waters, W., Gardner, N.: Test and analysis of buckling-critical stiffened metallic launch vehicle cylinders. In: 2018 AIAA/ASCE/AHS/ASC Structures, Structural Dynamics, and Materials Conference, Kissimmee, Florida, USA (2018)
16. Huybrechts, S., Tsai, S.W.: Analysis and behaviour of grid structures. Compos. Sci. Technol. **56**, 1001–1015 (1996)
17. Totaro, G.: Local buckling modelling of isogrid and anisogrid lattice cylindrical shells with triangular cells. Compos. Struct. **94**, 446–452 (2012)
18. Bednarz, A.: Evaluation of material data to the numerical strain-life analysis of the compressor blade subjected to resonance vibrations. Adv. Sci. Technol. Res. J. **14**(1), 184–190 (2020)
19. Quinn, D., Murphy, A., McEwan, W., Lemaitre, F.: Stiffened panel stability behaviour and performance gains with plate prismatic sub-stiffening. Thin-Walled Struct. **47**(12), 1457–1468 (2009)
20. Quinn, D., Murphy, A.: Non prismatic sub-stiffening for stiffened panels plates – stability behaviour and performance gains. Thin-Walled Struct. **48**(6), 401–413 (2010)
21. Peters, W., Ranson, W.: Digital imaging techniques in experimental stress analysis. Opt. Eng. **21**(3), 427–431 (1982)
22. Sutton, M., Orteu, J., Schreier, H.: Image Correlation for Shape, Motion and Deformation Measurements: Basic Concepts, Theory and Applications. Springer, Boston (2009)
23. Święch, Ł: Experimental and numerical studies of low-profile, triangular grid-stiffened plates subjected to shear load in the post-critical states of deformation. Materials **12**(22), 3699 (2019)
24. Arborcz, J.: Post-buckling behavior of structures. Numerical techniques for more complicated structures. In: Lecture Notes in Physics, vol. 228, pp. 83–142 (1985)
25. Bathe, K.J.: Finite Element Procedures. Prentice Hall, Upper Saddle River (1996)
26. Felippa, C.A.: Procedures for computer analysis of large nonlinear structural system in large engineering systems. by A. Wexler (ed.). Pergamon Press, London (1976)

Author Index

A. Mężyk et al. (Eds.): SMWM 2020, AISC 1336, p. 289, 2021.
https://doi.org/10.1007/978-3-030-68455-6

Printed in the United States
By Bookmasters